# CONFLICTS OVER NATURAL RESOURCES

A Reference Handbook

Other Titles in ABC-CLIO's
# CONTEMPORARY
# WORLD ISSUES
Series

Books in the Contemporary World Issues series address vital issues in today's society such as genetic engineering, pollution, and biodiversity. Written by professional writers, scholars, and nonacademic experts, these books are authoritative, clearly written, up-to-date, and objective. They provide a good starting point for research by high school and college students, scholars, and general readers as well as by legislators, businesspeople, activists, and others.

Each book, carefully organized and easy to use, contains an overview of the subject, a detailed chronology, biographical sketches, facts and data and/or documents and other primary-source material, a directory of organizations and agencies, annotated lists of print and nonprint resources, and an index.

Readers of books in the Contemporary World Issues series will find the information they need in order to have a better understanding of the social, political, environmental, and economic issues facing the world today.

# CONFLICTS OVER NATURAL RESOURCES

## A Reference Handbook

Jacqueline Vaughn

### CONTEMPORARY WORLD ISSUES

A B C  C L I O

Santa Barbara, California
Denver, Colorado
Oxford, England

333.7
Vau

10/11

Library of Congress Cataloging-in-Publication Data

Vaughn, Jacqueline.
  Conflicts over natural resources : a reference handbook / Jacqueline Vaughn.
    p. cm. — (Contemporary world issues)
  Includes bibliographical references and index.
  ISBN-13: 978-1-59884-015-5 (hard cover : alk. paper)
  ISBN-10: 1-59884-015-0 (hard cover : alk. paper)
  ISBN-13: 978-1-59884-016-2 (ebook)
  ISBN-10: 1-59884-016-9 (ebook)
  1. Conservation of natural resources—Handbooks, manuals, etc. 2. Natural resources—Environmental aspects—Handbooks, manuals, etc. I. Title.

  S944.V38 2007
  333.7--dc22
                          2006030998

ISBN-13: 978-1-59884-015-5          (ebook) 978-1-59884-016-2
ISBN-10: 1-59884-015-0              (ebook) 1-59884-016-9

11  10  09  08  07   10 9 8 7 6 5 4 3 2 1

ABC-CLIO, Inc.
130 Cremona Drive, P.O. Box 1911
Santa Barbara, California 93116-1911

This book is also available on the World Wide Web as an ebook.
Visit www.abc-clio.com for details.

This book is printed on acid-free paper ∞
Manufactured in the United States of America

*To Don Reisman,*
*for discovering great talent*

# Contents

# Preface

The United States has a heritage in both the preservation and the utilization of natural resources: its trees, water, minerals, and public lands. From the beginning of the country's history until today, efforts have been made to consider the effects of industrialization, westward expansion, land acquisition and disposal, population growth, and consumerism of those resources, once thought to be almost limitless.

This book is designed to provide an overview of the disputes over natural resources that are at the center of both U.S. and global debates. Chapter 1 begins with a more detailed definition of renewable and nonrenewable resources and identifies six key periods in world and U.S. history that have both defined and contributed to contemporary conflicts. Chapter 2 consists of four subsections that describe key controversies over natural resources in the United States, chosen from a much lengthier list because they are enduring issues. The first subsection, on minerals, oil and natural gas exploration and drilling, represents one of the most historically contentious bases of natural resource conflict. The battle over "black gold" begins with the discovery of deposits in the West and attempts to extract oil and natural gas to fuel the growing demands for energy by both consumers and industry.

In 1872, when land to create Yellowstone National Park was set aside by the government, the United States started a lengthy debate over protected area policy, the second subsection in Chapter 2. From the enactment of the Forest Reserve Act of 1891 to today's concerns about user fees and the resources of the National Park Service to maintain hundreds of designated units, stakeholders have argued over the cost of protecting scenic areas and wilderness, the resources available and how to increase them, and which sites should be protected.

Historically, public lands have been used for grazing livestock, with little attention paid to the ecological damage done to natural resources, such as soil compaction, watershed pollution, forest degradation, and the introduction of exotic species. Conflicts over grazing management, the third subsection in this chapter, involve ranchers and livestock producers, who feel that they have a right to use the land, and environmental organizations, which differ among themselves in their approach to protecting and restoring native biodiversity and self-sustaining ecosystems.

Lastly, Chapter 2 explores how the U.S. government has treated forests as a commodity, giving preference to large logging companies and the implementation of policies that have required the harvest of a continuous supply of wood for the growing nation. Those policies have led to litigation and protests over maintaining a sustainable supply of timber and, more recently, the protection of old growth forests.

Chapter 3 shifts the discussion of natural resource conflicts to the global level. Many of the disputes are localized or regional, even though they have international implications. The chapter begins with the first of four subsections, starting with conflicts over diamonds and precious metals. Gems, gold, and other minerals have been the source of conflict, especially in Africa, since the early twentieth century, fueling civil war among rebels, governments, indigenous peoples, and large trading companies.

As is the case in the United States, petroleum-producing countries are the sites of various types of unrest, ranging from the environmental damage caused by extraction and boundary disputes over oil reserves to military intervention to protect valuable oil and gas resources. The second subsection shows how increasing demand for energy makes conflict more likely and, increasingly, more violent. Some argue that the discovery of energy resources in Latin America and Africa leads to increased economic gain for all; others feel that development brings only hardship, corruption by ruling elites, and environmental disaster.

Chapter 3 continues with a discussion of deforestation in a global context. Human rights and environmental groups believe that timber exploitation and forest destruction have led to regional instability, civil war, and severe economic and social consequences for indigenous populations. Although the United Nations has sent peacekeeping missions to numerous countries, and imposed trade embargoes in some cases, nongovernmental organizations believe that resolution to these conflicts lies in better border controls and enhanced law enforcement.

The chapter ends with one of the most difficult natural resource issues facing the world today—water scarcity. Many of the conflicts stem from disputes among countries that share transboundary water reserves. An estimated fifty countries on five continents share reservoirs, rivers, and underground aquifers, and some experts believe that the next major world conflict will be over water.

Chapter 4 is a timeline of key events in chronological order that briefly annotates benchmarks in the identification of conflicts and how they have progressed over time, both in the United States and worldwide. Chapter 5 presents a series of biographical sketches, or profiles, of some of the individuals who have had an impact on natural resource conflict, either as participants in the disputes themselves or in resolving them. Some are well-known names, while others may be famous only in their home regions. Chapter 6 provides a selection of legislation, reports, resolutions, facts, and documents that provide historic background and a more direct presentation of conflicts and their resolutions.

Chapters 7 and 8 are designed to be used as resources for further study and discussion. Chapter 7 is an annotated list of organizations, associations, and agencies that relate specifically to the topics presented in Chapters 1, 2, and 3. The chapter includes contact information, including websites, for a broad range of groups. Chapter 8 follows with an annotated bibliography representing an array of print and electronic sources, from books and articles to videos and DVDs. These reference materials are generally available in public or educational libraries, making them accessible to a wide audience. Lastly, the book ends with a glossary of key terms that can be used to reference the other chapters as needed.

Since many natural resource conflicts overlap, it is difficult to cover them all in a single volume. But this book provides a generous overview that will allow the reader to develop both background knowledge and an awareness of specific issues.

# 1

# Background and History

The definition of the term *natural resources* is largely dependent upon the context in which it is used. Sometimes the phrase refers to renewable resources—those that can be replaced, either by human intervention or by nature itself. This usually includes trees and forests, water, wildlife, wilderness, soil, grassland and prairies, and marine life. Some definitions include nonrenewable natural resources—petroleum, natural gas, and hard minerals. One difference relates to the length of time it takes for resources to be replaced. Renewable resources such as trees can be replanted, fish habitats can be restored, and some wildlife can be reintroduced into their natural habitat. Nonrenewable resources like coal are limited, inasmuch as it has taken eons for them to be formed, and once depleted, they may never be available again. Other definitions refer to the fact that most natural resources are managed by the same network of governmental agencies and, therefore, ought to be combined. There is agreement, though, that the definition does not include problems relating to pollution, defined as the injection of a class of products into an ecosystem (water pollution and air pollution, for instance).

One way of working with such complex definitions is to begin with an overview of the history of natural resource management, by tracing the development of policies and laws that govern their use. Because the major issues that have arisen in the last century or so come from legal interpretations, legislation, and changes in public attitudes, a chronological approach is an appropriate way to explore the background of natural resource conflicts. Chapter 4 offers a more comprehensive timetable of key

1

events, which add detail to this chapter. Land management is key to understanding natural resources, because from most perspectives, land includes resources native to the soil, including trees and minerals. In more recent times, land law has expanded to cover activities taking place below and above the surface, such as drilling for oil, grazing cattle, and logging.

This chapter begins with the earliest records of human awareness about natural resources, which date from biblical times. The development of early English laws is the basis for policies and disputes that reach colonial America between the 1600s and the late eighteenth century, when the United States changes from a confederated system of government to a federal one. After that initial period of governmental involvement, researchers have identified six historical eras of public land management that form the basis for natural resource policy and conflicts today.

1. Acquisition (1781–1867): The public domain grew through a variety of methods: the addition of all lands from the Atlantic to the Pacific Ocean through the ceding of colonial lands to the federal government; purchase (the Louisiana Territory, Alaska); annexation (Texas); or negotiation (the Oregon Compromise).

2. Disposal (1781– ): The federal government has continually tried to balance its role as steward of the public lands and as property owner. Grants were made to homesteaders as a way of encouraging westward expansion; settlers were given land if they promised to plant trees, farm, or irrigate their property; veterans were "paid" in acres rather than in currency; new states were given land (including swamplands) that they could later sell to support education or transportation; railroads gained acreage alongside tracks that were being built; and direct sales were made to the public, a process that continues today.

3. Reservation (1872– ): The Constitution grants Congress the right to withdraw a specific portion of public lands from the public domain (the total inventory of government-held property) for future use, ranging from logging and mining to preservation. This power was used to create Yellowstone National Park, the national forest reserves, and various unit designations that would be administered by the National Park Service.

The Antiquities Act of 1906 gave similar power to the president for the designation of historic landmarks and other sites.

4. Custodial management (1897–1950): A series of statutes, starting with the Forest Management Act of 1897, marked the federal government's formal acknowledgment of its role as the caretaker of reserved lands. New agencies were established to administer public lands, such as the General Land Office, the National Park Service, the U.S. Forest Service, and the Bureau of Land Management. Over the years, as conflicts over natural resources developed, the agencies would change leadership, culture, and jurisdiction.

5. Intensive management (1950–1960): Trees have always been treated as a commodity, from the time when the Forest Service was included under the umbrella of the U.S. Department of Agriculture. The cutting of timber produced income, jobs, and needed lumber for post–World War II development. Leases were granted for oil and gas exploration, grazing privileges, mining operations, and even recreational use.

6. Consultation and conflict (1960– ): The use of public land became the source of major conflict just as the environmental movement was reaching its prime. The concept of multiple use (designating the types of activities that could be conducted, from skiing to wilderness preservation to extraction) became the basis for natural resource management. Congress became more involved in decisions about land use, agencies battled over the coordination of conflicting activities, litigation became a key environmental group tactic, and the public sought a greater participatory role in decision-making. Budgetary considerations often shaped policy more than scientific research, a problem that continues to cloud key issues described in Chapter 2 (Clawson 1983).

It is important to note that there is some disagreement about these somewhat arbitrary categorizations of time. Certainly, conflicts relating to timber management in the United States have been virtually continual since 1691; the Charter of Massachusetts contained a provision protecting trees for the masts of ships for the

Royal Navy. Today, similar disputes divide the federal government and environmental organizations over the practice of clear-cutting and the management of old-growth forests.

But natural resource policies and the conflicts arising from them have changed over time, and there are specific periods when disputes have focused on the actions of a presidential administration or an individual statute. The following sections should be viewed as just one way of explaining historical trends and key events.

# From the Dawn of History

Identifying the point in time when concern about natural resources developed is an imprecise science. Some point to scriptural references in the Bible that mention the earth's bounty and humankind's responsibility for stewardship over its creatures, a reference found in the Book of Genesis. There is mention of lands in which nothing is lacking, and also of human exploitation, in the book of Deuteronomy. Verses and their meaning are always subject to interpretation, of course, and not everyone agrees on simple issues such as the accuracy of translation from ancient languages to modern ones. The concept of stewardship has re-entered contemporary religious dialogue, however, as a way of bridging Scripture and modern environmental challenges such as energy use and water scarcity.

By the fifth century, Christianity was the predominant faith in much of the Roman Empire, and by a century later, most western Europeans had converted from paganism. Paganism was based on the premise that people and nature share a spiritual connection, and that spirits are present in all forms: trees, rivers, mountains, and as elves and fairies. Paganism called upon its followers to respect the natural spirits and to offer prayers before killing an animal or cutting a tree. Christianity, from about A.D. 500 to 1500, was more likely to promote the dominion of man over nature, and the harnessing of resources became God's gift to man (Kline 2000).

Many legal scholars believe that contemporary views about natural resources stem from Roman law and the customs and traditions that would later be systematically compiled into codes governing behavior and property during the fifth century A.D. The *Digest of Justinian* and the *Institutes of Justinian*, named for the ruling emperor of the time, contained specific reference to common property resources, or *res communes,* which were owned by and open to everyone (Adams 1993).

During the Middle Ages, there was general agreement that forests were to remain wild, with little attention paid to their value for production. There were separate forest courts and a complicated hierarchy of officials responsible for the conditions of the forests, enforcing grazing regulations and monitoring the availability of game for hunting. Laws governed the clearance of land, the cutting of tree limbs, and the building of ponds. This led to conflicts between the monarchy and freeholders (landowners) as land was reserved for royal forests through unofficial routes of acquisition. By A.D. 1217 the disputes, driven largely by the economic impact of the Crusades and foreign wars, led to revisions in forest law and the issuance of the *Carta de Foresta,* which would later become part of the Magna Carta. While the codification of forest law could be seen as benefiting some, lands that had previously been protected quickly became open to timber cutting, especially during the fourteenth and fifteenth centuries, when wood was needed for ships for trade and exploration. What may have started as land reform ended up as mismanagement driven by greed and corruption (ibid.).

# Colonial America

Early settlers found that Europeans expected the New World to be a source of materials to meet their needs, whether spices, timber, or meat. There was a kind of grocery-store mentality that left the impression among some that America was a storehouse that could be raided as often as necessary. What is more important, colonists imported laws and customs that governed land ownership, something foreign to the native tribes that had lived there for centuries before European settlement. The idea of buying and selling land, which belonged to no single individual or family (partly because of the mobile nature of much of the Native American population), was based on the English system. In the early-seventeenth-century settlements, lands and resources were treated much the same as the feudal relationships of monarchs and their subjects. The privilege of land ownership was extended primarily to the upper classes, granted by a king to reward military service, as a reward for loyalty, or through birthright.

By the mid-1700s, the abundance of land encouraged some settlers to move to unclaimed land that had not been surveyed; squatters simply declared ownership, legal or otherwise. If a parcel of land was not claimed by a white man, then it was considered free for the taking. Native Americans who had occupied the land were driven off or killed so that colonists could take what they believed was rightfully theirs.

In the Virginia Colony, men received a warrant for an area from the land office, found acreage they wanted, defined its boundaries using crude maps, and made their claim (called a patent). After the Revolutionary War, colonial leaders expanded their claims westward, with boundaries often overlapping and large grants of lands made to influential individuals. To reduce conflicts over boundaries and land holdings, the survey process was formalized, with some land auctioned off for minimal prices to the highest bidder. By this time, land ownership was a firmly entrenched custom, even though it meant that native populations "lost" something they never had considered as theirs to begin with.

Among the earliest natural resource disputes were conflicting claims by the colonies/states over the land between the Appalachian Mountains and the Mississippi River. The colonies of Connecticut, Georgia, Massachusetts, New York, North Carolina, South Carolina, and Virginia sought control of large tracts of land. The Maryland colony, which had no land claims in the disputed area to the west, argued that the territory had been secured through bloodshed against the British Crown and should be considered as "common stock." At stake, in addition to the western lands, was the signing of the Articles of Confederation. Maryland refused to sign the document, with leaders arguing that the land claims would put it in an unfair economic and political position.

To resolve the impasse, New York agreed to cede its claims on the western lands in 1780; by 1802, the remaining states had done so as well. Maryland signed the Articles of Confederation, and agreement was reached that the newly formed Continental Congress would be responsible for any future land disposal. As new territories were added, any lands that were not part of existing claims by corporations or individuals remained under public ownership (Dombeck, Wood, and Williams 2003). These political agreements appeared to resolve the question over what to do with the new frontier, although the debate was really just beginning.

# An Overview of Natural Resource Issues and Conflicts

To understand contemporary natural resource conflicts, it is useful to look to the past and the development of U.S. policies relating to public lands, because that is where many of the current disputes originated. The issues that divide the country have not always been the same, but there are common themes that can be traced through each of the following six periods.

## Acquisition: 1781–1867

The state cessions, which took place between 1781 and 1802, increased the public domain by 236 million acres, at a cost of $6.2 million to the newly formed government. With control of most of the acreage between the Appalachians and the Mississippi River, an additional 10 percent of the total public land had been added. Although the early explorers and colonists had little idea what lay to the west of the Mississippi River, they did expect to find more land and some kinds of riches—at least based on rumors and stories passed along by explorers and Native Americans. The states and federal government now owned all of the land east of the Mississippi (except Florida, which was claimed by Spain), and there was an agreed-upon desire to add as much more as possible. Land meant wealth; the more land the government owned, the more credibility it developed and the more resources it was likely to find and control.

One of the driving forces behind the acquisition process was the concept of Manifest Destiny. In 1839, John L. O'Sullivan wrote in an article in *The United States Democratic Review* that Americans were unique in history because "our national birth was the beginning of a new history, the formation and progress of an untried political system, which separates us from the past and connects us with the future only; and so as regards the entire development of the natural rights of man, in moral, political, and national life, we may confidently assume that our country is destined to be the great nation of futurity." O'Sullivan went on to describe the boundless future of U.S. greatness, writing: "In its magnificent domain of space and time, the nation of many nations is destined to manifest to mankind the excellence of divine principles" (O'Sullivan 1839).

The vision that O'Sullivan described became the clarion call and campaign promise for dozens of political and social leaders during the 1840s. Manifest Destiny carried with it a sense of the need to fulfill a spiritually driven mission of converting "the immutable truth and beneficence of God...to the nations of the world, which are shut out from the life-giving light of truth" (ibid.). This was interpreted to mean that it was America's duty to spread democratic principles of freedom and brotherhood to the rest of the continent, and eventually to the world. It, by inference and later policy, excluded those incapable of self-government, including non-Europeans and Native Americans, who were believed to be savages. Purchasing, annexing, or taking the land east of the Mississippi River was considered the mechanism for fulfilling that destiny.

Before O'Sullivan wrote about Manifest Destiny, the process of acquiring land for the new nation had already begun. In 1803, President Thomas Jefferson led the acquisition rush with the Louisiana Purchase: 529 million acres of land between the Mississippi River and the Rocky Mountains, or about 23 percent of the public domain. The $15 million agreement with France was considered by some an extravagance, but it typified Jefferson's thirst for knowledge, which could be quenched only by finding out what was there. It also established a firm economic base that Jefferson believed would make the country self-sufficient.

Almost immediately after securing the purchase, Jefferson arranged political and financial support for the exploration of the West in hope of finding a trade route to the Pacific (among other resources). From 1804 to 1806 the Lewis and Clark expedition traveled the waters of the frontier, crossing the Rocky Mountains and meeting native people, some of whom were friendly and some of whom were not. After returning to Washington to report to Jefferson on the sights they had viewed in their long journey, the expedition's members were unable to grasp, or convey to the president, the vastness of the Western frontier. Although Jefferson was considered a visionary, he thought it would take a thousand years for the region to be reached and settled (Kline 2000).

The Louisiana Purchase was followed by acquisition of the 30 million acres of the Red River Basin (1782–1817), and the addition of Florida (a cession from Spain in 1819 that added 46 million more acres for $6 million). At this point, the United States was already experiencing both a high birthrate and immigration that led to a population boom. Having large families to work the growing agricultural areas of the nation's midwest was considered not only desirable but also necessary. Expansion into new territories was needed to provide more land for the country's 20 million residents, 4 million of whom moved to the Western territories between 1820 and 1850. Economic depressions in 1818 and again in 1839 forced many urban dwellers and farmers to head toward the frontier areas, which were still being explored, opening up new opportunities for land ownership, commerce, and self-sufficiency. A new phase of land acquisition resulted from negotiations with Great Britain that resulted in the Oregon Compromise of 1846 and 183 million acres added to the public domain.

Mexico, which had achieved its independence from Spain in 1821, faced a different situation. Internal political struggles left the country in financial and social ruin; attempts repeatedly were made to colonize its borders despite fierce battles with Native Americans. The frontier communities were lawless and poor despite the efforts of the Catholic Church and military to civilize the natives. The United States took advantage of Mexico's situation in the 1848 Mexican Cession, adding 389 million acres as conditions under a peace treaty and boundary dispute. Texas, which had won a war of independence with Mexico and had operated under independent rule for nine years, was annexed in 1845; an additional $10 million was spent to acquire 19 million acres of land along the current U.S.-Mexican border which became Arizona (the 1853 Gadsden Purchase).

Among the most controversial land acquisitions of all was the 1867 Alaska Purchase, also known as Seward's Folly. The Russian government, which had held claim to the Alaska territory since 1741, offered to sell it to the United States because of repeated land disputes between U.S. and British settlers. More important, the Russian czar needed money, so in December 1866, the Russian minister to the United States was told to negotiate the sale of the territorial lands. The secretary of state, William H. Seward, initially offered to purchase 365 million acres for $5 million in gold, an amount that the Russian minister, Baron Edouard de Stoeckl, was authorized to accept. But believing that Seward's dreams of expansion would not allow him to walk away from a deal, Baron de Stoeckl negotiated a better deal, and Seward ended up agreeing to pay $7.2 million, or about 2 cents per acre—a bargain still, but criticized heavily at the time. Seward persuaded President Andrew Johnson to call the Senate back into special session to approve the sale and treaty, which radical Republicans derided because the government had bought "the northern ice-box." Although the Senate finally approved the agreement on April 9, 1867, the House would not appropriate the necessary funds to complete the purchase. After numerous bribes, scandals, and the impeachment of President Johnson, the purchase of Alaska was finalized in July 1867. The added land was twice as large as Texas, but most Americans saw little value in Alaska until the Great Klondike Gold Rush in 1896. In sixty-four years the United States had grown from the original 500 million acres along the East Coast to more than 2.3 billion acres, spanning all of the land between the Atlantic and Pacific oceans. Manifest Destiny appeared to be complete.

## Disposal: 1781–

Despite the flurry of acquisition, the new government quickly realized two things: it did not want to be in the public land business, and there was an overwhelming need to produce revenue not only to run the country but also to pay off debts incurred during the Revolutionary War. Some political leaders saw the newly purchased public lands only as a commodity to be sold as a source of revenue. Many Easterners wanted the "new" land conveyed to individuals, while some, such as Daniel Webster, believed that the country's first priority was to get the frontier settled and economically productive. In 1783 the population of the United States was about 3.25 million, about one-third of whom were slaves or in jail. By 1850 the total population had reached 50 million, and between 1851 and 1860, 2.6 million immigrants, mostly from Ireland and Germany, arrived to settle the country and to become the major source of cheap labor.

A number of factors eventually led to the passage of the General Land Ordinance of 1785, one of the first major signs of a change in land policy. Among other provisions, Congress granted public lands to the states to support education, based on the formula of one section of land per township. These "trust" lands were to provide states with a revenue source (land that could be sold or leased as needed) and would prioritize the provision of education, especially in the newly emerging West. In 1850 the school land grants were increased to two sections per township, and Utah, New Mexico, and Arizona received four sections per township. This form of land disposal accounts, in part, for the high percentage of state-owned lands in those three states, although the federal government was still the majority landowner. The existing states also protested the efforts by Western territories to convince the federal government to give them public land, arguing that the public domain benefited the new nation and should be held in reserve for future use. As a condition of entrance, new states agreed not to tax federal property or to interfere with the disposal of federal lands. They eventually accepted agreements granting them money and other concessions instead.

Over time, states began to interpret the 1785 statute in different ways. Some specifically required that sales from trust lands be used for infrastructure development such as new roads or public buildings. Other states, such as Idaho, constitutionally required that state lands be managed for the highest return, which usually meant that the land be sold to the public or leased for single-purpose harvest, such as mineral exploration or logging (Cawley 1993).

The government encouraged the rush to the West, supporting settlement by giving land away. The 1862 Homestead Act granted 160 acres of land to any settler who agreed to live on it for a minimum of five years, developing it through farming or ranching or some other type of activity that "controlled" nature. The domestication of the wilderness, as it has been called, was based on the idea that the European garden could be easily transported to even the most arid and desolate areas of the West. One settler, Henderson Luelling, an Iowa Quaker, hauled 700 trees, vines, and shrubs along the treacherous Oregon Trail in 1847. The fruit trees and other plants brought by pioneers in the Pacific Northwest would become the basis for the multimillion-dollar orchard and fruit industry in the region (Kline 2000).

In some cases, when settlers encountered large groups of Native Americans on land they wanted, they attempted to deal with them peacefully by trading goods for land. Native Americans were valuable to settlers for their knowledge of the area and its resources, and whites offered alcohol for maps and trapping techniques, colorful beads and fabric for buffalo robes. Unlike the indigenous population's understanding of the ecology of native plants and animals and their role in maintaining a steady supply of food, whites wasted many of the natural resources they considered to be in abundance. Trees were burned down rather than felled, and wild animals like squirrels that found their way into granaries were considered pests subject to bounty hunting. European and eastern women eagerly purchased fur coats and hats and added the plumage of exotic birds to their hats as fast as they could be killed.

Shortly after the government had reached the Pacific Ocean in its move for acquisition of new land, divestiture became the predominant federal policy, encouraging speculation, fraud, bribery, and, over the next several decades, the kind of environmental misuse and degradation that became the basis for the next phase of natural resource policy, the creation of federal reserves.

## Reservation: 1872–

The concepts of conservation, preservation, and reservation are closely intertwined in the history of natural resources. While it would be simple to attribute those ideas to influential writers such as George Perkins Marsh, Henry David Thoreau, and Ralph Waldo Emerson, that was not always the case. In 1832, Congress set aside four sections of Hot Springs, Arkansas, for "future disposal," with the idea that they might be developed at a later date (the land later became part of the national park system). In 1864, Senator John Conness of California sponsored legislation that granted the Mariposa Grove of giant redwood trees to the state for "public use, resort, and recreation," to be held "inalienable for all time" (Adams 1993, 179). Conness was not known as a preservationist; the request for the legislation had come from a representative of the Central American Steamship Transit Company and the manager of the nearby Mariposa Mining Estate. The idea of setting aside land for future use became more common after the passage of legislation that set aside public land in the Yellowstone area in 1872. The statute made clear that the government had the power to "reserve" some of the public domain for specific purposes, an idea that was clarified by other acts of Congress over the next thirty years. The Forest Reserve Act of 1891 (also known as the Creative Act) gave the president the power to set apart and reserve public timberland as reservations. The 1891 statute is considered a milestone in the history of U.S. environmental law because it not only became the basis for the development of forest policy but also served as the impetus for the creation of protected areas, such as the national parks.

The law marked one other major change. The federal government stopped the disposal of public forest lands into private ownership, and in 1911, under the provisions of the Weeks Act, the Forest Service was authorized to purchase forests that had previously been owned by individuals and timber companies. The land, in many cases, was nearly barren, all of the marketable timber having been harvested, and many owners stopped paying taxes on their property, simply abandoning it. Thousands of acres west of the Mississippi were purchased and gradually restored, expanding the federal government's timber resources.

The power to reserve large tracts of land was disputed by economic interests, however, especially large lumber and mining companies. Along with ranchers and farmers, they felt that the federal government had exceeded its authority, tying up the land and giving them limited access to the resources on those lands— timber, oil, minerals, and grazing areas. The government sought to limit private exploitation of natural resources, and the extractive industries sought to have the lands in question returned to the states for control.

Unlike the lands that were reserved for protected areas and for timber, the government did not secure the same type of protection for mining lands. The 1872 Mining Law emerged after gold was discovered in California; it represents one of the few areas where natural resource conflicts were initially quite minimal. Although there were disputes between miners over individual claims, the federal and state governments seemed content to legitimize claims with little interference. Since no bureaucracy existed at the time to regulate mining, the only administrative control was provided by the U.S. Army. The early mining industry basically worked out its own system of property rights and privatized any form of dispute resolution.

Less than a month after the Forest Reserve Act was signed in 1891, President Benjamin Harrison used his power under the law to designate the Yellowstone Park Forest Reservation, expanding the size of the initial park. In 1906, Congress passed the Antiquities Act and delegated authority to the president to designate historic landmarks and objects and sites of historic or scientific interest on federally owned or controlled land. The concept of reservation further protected valuable watersheds, areas of scenic beauty, and forest lands for future use or enjoyment. By 1900, there were six national park units covering more than 4 million acres; by 1916 there were twenty national monuments, many of which would later become the centerpieces for future national parks.

But at the same time, there were few resources available to manage the reserved areas. Many of the park superintendents served without pay or had no budget, and the reserve staff often consisted of army soldiers who were authorized to prevent trespassing and poaching. Later, army engineers would be used to build roads and bridges and to mark park boundaries, and cavalry troops provided patrols and protection. The management of the reserves and monuments was fragmented and controlled by three federal agencies: the departments of Agriculture, Interior, and War.

Two individuals, Gifford Pinchot and John Muir, epitomized the very public debates over how natural resources on public lands should be used. Pinchot, who had been trained in forest management in Europe, and who is considered the first American forester, believed that "the rational management of forests could not be successfully conducted without 'the supervision of some imperishable guardian; or in other words, of the state'" (Miller 2001, 94). Governmental control of forests would establish a base for scientific forestry on the ground, he said. Those principles were dependent upon two ideas, Pinchot believed: that individual trees required many years before they reached what he called "merchantable size," and that a "forest crop" could not be harvested year after year from the same plot of land. Forestry practitioners must be patient, he said, and the government needed to take a paternal approach to forest lands.

Pinchot's vision of conservation was based on three principles: development (the use of natural resources for the benefit of people here today); prevention of waste; and the development and preservation of natural resources for the benefit of the many, and not merely for the profit of the few. Those principles, he believed, would result in planned, orderly development, and eventually national efficiency. Eventually, this view would become known as utilitarian conservation.

John Muir, in contrast, had a different view of resource use. Although Muir and Pinchot often traveled together and frequently socialized, Muir sought the preservation of wilderness lands, not enhanced use and commercial development. Muir thought of himself as a mentor to Pinchot, and agreed that national forests should be both preserved and used. He supported Pinchot through the years of the late nineteenth century because he believed that scientific management was preferable to past practices that obliterated forests.

But around 1898 their views diverged sharply, resulting in one of the most bitter conflicts over natural resources in the his-

tory of the United States. The issue that divided them was a proposal, initially made in the 1880s, that the Hetch Hetchy Valley in
Yosemite National Park be dammed to create a reservoir for water
that would serve the residents of San Francisco. In 1903 and 1905,
the city applied to the Department of the Interior for a permit to
build the dam. The permit was denied on the basis of the belief
that it would damage the spirit of the national park. Pinchot, who
was head of the U.S. Forest Service in the Department of Agriculture, told Interior Department officials that the dam would not
detract from the beauty of Yosemite, and he argued that there
needed to be a tradeoff between preserving Hetch Hetchy and the
needs of the communities in the growing San Francisco area.
Muir, in turn, urged Pinchot to ignore "the benevolent outcry for
pure water for the dear people," and warned of graft and corruption if the dam were to be built. He countered with a plan to build
a dam below the valley (Miller 2001).

The 1906 San Francisco earthquake led to fires and the break-
down of the city's water system, and city officials reapplied for
permission to build in Hetch Hetchy, believing that with Pinchot's support and that of a new secretary of the Interior, they
would finally be successful. But Muir, who had been one of the
founders of the Sierra Club in 1892, marshaled his friends and
successfully lobbied Congress to oppose the dam. When the two
men met in 1907, Pinchot admitted to Muir that he had never seen
the Hetch Hetchy Valley and was unaware of how it was an important part of Yosemite National Park. He told Muir to send a
letter to the secretary of the Department of the Interior asking that
no action be taken until the Sierra Club's views could be considered. Muir thought that Pinchot was becoming less adamant
about the dam, but, in fact, Pinchot wrote President Theodore
Roosevelt that the highest possible use of the Hetch Hetchy
would be to provide a source of water for San Francisco.

The question of the dam was revived in 1913, when Congress
took up the issue in hearings in which both Pinchot and Muir testified. According to Pinchot, preservation of the valley would
benefit only a small group of elites, compared with the benefits to
thousands of residents. Muir countered that the dam would actually benefit special interest groups and was a political ploy to secure the political ambition of the city's leaders. The eventual decision to build the dam was made in 1913 by President Woodrow
Wilson, whose secretary of the Interior had previously served as
the city attorney for the city of San Francisco.

Hetch Hetchy symbolizes the tug-of-war between conservation and preservation that would continue well into the twentieth century. However, it also shows how Pinchot's vision of conservation became the nation's resource management philosophy for the next fifty years. Commodity users became an important political force, dominating the development of new legislation and policies over how natural resources would be used until the birth of the contemporary environmental movement in the 1960s.

## Custodial Management: 1897–1950

Just before the turn of the century, the federal government shifted its role from simply owning land to taking a more active role in managing it, including its natural resources. The Progressive movement, conservationism, scientific management, and the emphasis on efficiency were gradually affecting both the government and professions dealing with resources. The generic approach to land was broken down into its organic parts, with laws, agencies, and technicians divided by resource. Management by expertise became the norm, with foresters determining how much timber should be cut in a specific plot, and agronomists making decisions about the number of cattle that could be grazed and how much water was needed.

Custodial management was intertwined with the slowly emerging discipline of public administration, and the development of what has been called the resource management state. The term refers to the replacement of a legalistic, localistic system with a centralized, bureaucratic one. During the presidency of Theodore Roosevelt, the United States became a frontrunner in environmental policy. As one historian notes: "Assembling a cadre of trained, dedicated public employees, resource management agencies initiated policies, introduced new techniques of public management, and literally reshaped the nation's landscape" (Schulman 2005, 378).

The structure and reorganization of the federal government was another element of custodial management. Legislation created new agencies, new policies, and new leaders. The jurisdictional disputes were based on securing authority over natural resource management. Pinchot had started his service in the federal government in 1898 as an administrator for the Agriculture Department's Division of Forestry, which at the time supervised all of the government's foresters. The federal forest reserves themselves, however, were under the jurisdiction of the Department of the Interior. Pinchot wanted authority over both the foresters and the forests, and after a long struggle he was successful. In 1905 his agency was renamed the U.S. Forest Service, the forest reserves became national forests, and Pinchot built up an institutional army of loyal employees who were compared to the U.S. Marines and totally devoted to their chief (Schulman 2005).

A similar phenomenon took place with the Bureau of Reclamation under the leadership of Frederick Newell, one of Pinchot's allies in Washington. The "age of the engineer" gave the bureau considerable credibility at a time when the country was deeply immersed in the building of dams, flood control systems, canals, and hydroelectric projects. The Department of the Interior, however, did not fare as well when federal reorganization efforts began under Roosevelt. Pinchot and Newell wanted to have all natural resources under the Department of Agriculture, arguing that Interior lacked the skilled employees, the necessary office equipment, and the rigorous organizational structure that was needed to manage public lands. Field offices were often staffed by patronage appointments and were greatly susceptible to fraud.

In 1905, Pinchot convinced Roosevelt to establish the Committee to Investigate the Executive Business of the Government, more commonly known as the Keep Commission after its chairman, Assistant Secretary of the Treasury Charles Keep. The commission, charged with a review of the executive branch, uncovered incompetence throughout the Department of the Interior, with a promotion and rewards system that promoted the most unskilled clerks to positions of responsibility. The Land Office was nothing more than an administrative nightmare, according to the commission: "The fact that possession of land has been ordained through perjury with perfect impunity for many years through many of the regions of the West has naturally resulted in complete public indifference to this particular crime. In addition, the laws and officers of the United States have become subjects of public ridicule, both from the glaring character of the frauds and the ease with which they have been accomplished" (Schulman 2005, 387).

Despite the Keep Commission's scathing indictment of the existing administrative structure for natural resource management, Congress rejected every one of the initiatives that required legislative approval, even refusing to fund the Keep Commission's official reports. When Roosevelt left office, President

William Howard Taft forced Pinchot and his supporters out of office, and the vision of the autonomous natural resource bureaucracy was left behind.

Pinchot's influence was not totally gone, however. The Hetch Hetchy controversy with Muir heightened public awareness about the importance of managing wilderness areas like Yellowstone and Yosemite, and led to the creation of the National Park Service in 1916. The dispute strengthened the notion that development ought to be kept out of the national parks in order to preserve their scenic beauty for future generations. The concept of stewardship was a dramatic change in how reserved areas were to be managed.

Preservation groups also emerged around the turn of the century, especially in the East. Organizations such as the National Audubon Society and local associations sought more designations, often aided by wealthy and influential businessmen. John D. Rockefeller provided financial assistance to a group formed to protect Mount Desert Island, off the coast of Maine. The New England preservationists began acquiring and purchasing about 6,000 acres that were threatened with development, and in 1916 they persuaded President Wilson to use the Antiquities Act to designate the area as a national monument.

The groups were often aided by the National Park Service itself, which created study commissions to determine whether there were appropriate sites for consideration as national parks. The Temple Commission, chaired by Pennsylvania congressman H. W. Temple, surveyed the southern Appalachian region in 1924, examining sites of at least 500 square miles that included mountain scenery, areas that could be used for visitor recreation, untouched forests and waterfalls, springs and streams for fishing, wildlife conservation, and perhaps most important, sites that were accessible by rail and road. The result of the commission's recommendations was a 1925 statute directing the secretary of the Interior to determine the boundaries of what would eventually become Shenandoah, Great Smoky Mountains, and Mammoth Cave national parks (Adams 1993, 192–193).

Problems of visitation were already beginning to plague several of the most popular national parks. In 1890, President Benjamin Harrison signed legislation creating Sequoia National Park in California, the home of giant, 2,000-year-old trees. The area initially needed to be protected because of damage being done by sheepherders, prospectors, and loggers who had begun to fell the valuable timber. By 1903 the first road was built, followed by automobiles that replaced pack trains, and by 1930 there were four campgrounds, more than 200 cabins, corrals, a gas station, retail stores, offices, and dining halls catering to tourists. Water and sewer systems were built, damaging the shallow roots of the trees, and buildings and parking lots weakened them even more. The park's superintendent called for commercial activity to be moved out of the Giant Forest area, but his recommendations were ignored. To make more space for visitors, during the 1950s the National Park Service (NPS) would move out cabins, cut down the trees, and then move the cabins back in (Leonard 2005).

Protected area policy became more and more controversial as the number of designated units and acreage grew. Lumber companies, for instance, were hostile to the idea that prime timberlands were being declared off limits. In 1907, Western members of Congress garnered sufficient opposition to the forest reserve concept that they took away the president's authority to create or enlarge forest reserves in six Western states. States were angry that they were expected to pay for the acquisition of park land, or to conduct fund-raising efforts to do so. The Great Depression led to cuts in agency budgets and subsequent reductions in park staffing and maintenance. Park protection was often limited, and management of the national forests was curtailed to minimal levels. Mining companies objected to the idea that they could not purchase land but could only lease it.

Salvation came in the form of the Civilian Conservation Corps (CCC), created in 1933 as part of President Franklin Roosevelt's New Deal program as a way of putting unemployed men to work, even at the most menial jobs and minimal pay. More than 250,000 men were immediately mobilized, paid $30 a month, given clothing, medical care, housing, and training, and sent to the 17,000 CCC camps that were created. Most were deployed to work in the Forest Service and became part of a massive reforestation and restoration effort. Many were assigned to timber management programs that probably would not have occurred otherwise, or to firefighting efforts. The CCC was terminated in 1942, but its legacy can still be found in trails built by young laborers, national park cabins and lodges, or in the initials they carved into bridges and roadways.

In 1940, the national park system included about 22 million acres of land for a population of 130 million people; by 1960 only a few more acres had been added, while the population grew to

183 million. The parks were overcrowded, and the NPS had a limited capability to manage the visitors, let alone the resources within the parks. While resource management had been a goal of the agency since its formation, the reality was that the demand for recreation outstripped staffing and budgets.

The management of rangelands underwent a similar transformation. The U.S. Forest Service held responsibility for grazing on public lands for the first thirty years of the twentieth century. Until 1906, there was only minimal support for any type of regulation of livestock grazing on the nation's vast grasslands and prairies. The doctrine of custom and culture, explored more in Chapter 2, dominated the West, and ranchers seldom encountered any resistance to moving their cattle or sheep to more productive land, whether public or private. The General Land Office, charged with regulating livestock access on public lands, was replaced by the Grazing Service, and in 1934 by the Taylor Grazing Act. Under the new law, ranchers who had previously taken their herds wherever they wanted now had to apply for federal permits and to consult with local grazing advisory boards. For an industry that was used to unfettered access to the land, these types of regulatory policies were unfathomable.

While the government's goal had been to create economic stability within the ranching interests by developing a more structured management approach, any structure at all was certain to be opposed by ranchers. To counter the initial resistance to the 1934 law, the government agreed to keep grazing fees artificially low and to give existing grazing interests priority in the permitting process. Attempts to increase fees were met with strong congressional opposition, and the eventual creation of the rancher-friendly Bureau of Land Management (BLM). Influential stockmen's associations supported Western members of Congress, who held the BLM's purse strings, creating a decentralized management structure that was controlled by local-level interests.

Mining regulation remained virtually intact after the passage of the 1872 Mining Law. It was not until after the turn of the century that the government took particular interest in hardrock minerals, oil, gas, and chemicals. About 150 million acres of public land were gradually reserved as the conservation ethic began to emerge, and the public sought to have some sort of return for the profits being made by large companies. In 1917, Congress granted the secretary of agriculture authority to lease lands with hardrock mineral claims on them, and in 1920 the Mineral Leasing Act was passed. In order to maintain some level of government control, the statute allowed private companies to lease land, but not to purchase claims. Instead, a rudimentary royalty system was passed to give the federal government a percentage for resources found on public land, a program that did not invoke controversy until the advent of the environmental movement.

## Intensive Management: 1950–1960

After World War II the United States changed dramatically, from shortages to abundance, from a work-intensive society to one that had leisure time, and from war-related industries to more specialized, technology-driven companies. Americans had more disposable income, more time in which to spend it, and more transportation options that gave them a sense of almost limitless mobility. These changes opened the door to recreational activities and created a desire for families to enjoy the outdoors.

At the same time, the U.S. Forest Service was attempting to manage the national forests under conflicting mandates. The 1897 Organic Act required the agency to improve and protect the forest reserves, while at the same time requiring the Forest Service to provide a continuous supply of timber for the nation's needs. There were competing goals of protection, regulation, and public use that led to decades of conflict both within the agency and outside of government.

In parallel actions involving the BLM and the Forest Service, the federal government began to take a second look at public rangelands. Through the 1950s and early 1960s, ranching interests still controlled grazing policies despite any attempt to weaken their hold on the BLM. Groups such as the National Cattlemen's Beef Association, which had developed statewide chapters, continued to make large contributions to Western legislators and joined with other powerful interests whenever any change in public land use was being proposed.

Few of the early environmental organizations focused on energy policy, concentrating instead on wilderness and natural resources. Ironically, the postwar period's emphasis on consumerism and mobility was having a major impact on the oil and gas industry. More vehicles on the road meant a higher demand for fuel, as did consumer purchasing of appliances and overall national energy needs. Congress, however, was hesitant to initiate any major changes in energy laws, and it limited its regulatory activity to three pieces of legislation. The Multiple Mineral Development Act of 1954 covered hardrock minerals claims being leased for oil, and the Common Varieties Act of 1955 deleted provisions related to substances such as pumice, clay, and cinders from the 1872 Mining Law. Another 1955 statute, the Surfaces Resources Act, restricted the surface use of mineral claims to those required for mining (Davis 2001). Generally, though, consumer interests for more energy trumped environmental organizations' efforts to reform the 1872 law. Most groups were still focused on pollution and wilderness, and the petroleum and mining industries' power in Washington stayed strong enough to keep conflict to a minimum.

## Consultation and Conflict: 1960–

Some scholars believe that one of the major factors influencing natural resource conflicts since 1960 has been the country's demographic changes. In 1960 the population of the thirteen Western states was about 28 million, or 15.7 percent of the nation's total. Over the next forty-five years, not only would the percentage of Western residents increase substantially but, in addition, the region would become the most urbanized area of the United States. In 1990, about three-fourths of the nation was considered urbanized; in the West, more than 86 percent was urbanized, with nearly 89 percent of the Pacific considered urban (ibid.). The most dramatic growth has been in the states of Alaska, Arizona, and Nevada—states that also have a high percentage of public lands.

The change from a rural population to an urban one has brought about changes and conflicts over natural resources. Urbanization has led to a reduced reliance upon mining, logging, and other extractive industries, one of the factors responsible for the demise of many timber- and mining industry–based small towns. As jobs in those industries have disappeared, so, too, have the people who once depended upon them. Not everyone left communities like Superior, Arizona, or Roseburg, Oregon, but younger generations found that the loss of a major employer also meant the loss of opportunity.

With the publication of Rachel Carson's groundbreaking book, *Silent Spring*, in 1962, the public gained an awareness of the dangers posed by many of the chemicals that had been developed during World War II. Pesticides, chemical fertilizers, and toxic substances that had been used as part of the military arsenal had found their way into U.S. agriculture. Carson documented the impact on the country's songbird populations, noting that she no longer heard the bird calls that usually accompanied the start of spring and the end of hibernation. Instead, the avian world was silent, and she had decided to find out why. Carson's research led to a series of warnings and debates over the role of technology, and questions about why so many substances had been used with so little concern about how they might affect the soil, plants, wildlife, and human beings. She was vilified by the chemical industry; many scientists criticized her research methods, motivation, and even the fact that she was a woman. A Federal Pest Control Review member said that he "thought she was a spinster, [so] what's she so worried about genetics for?" (Kline 2000, 75).

*Silent Spring's* influence went far beyond alerting the public to the dangers of chemicals, however. It opened up a period of citizen participation and activism that resulted in the beginnings of the contemporary environmental movement. That activism quickly spread from initiatives related to toxic chemicals to concerns about forests, natural scenic areas, wilderness, and other natural resources. What had been started by Muir and the Sierra Club expanded during the 1960s and 1970s to cover every conceivable species and issue. President John F. Kennedy pushed forward pollution legislation during his brief administration, and his successor, Lyndon Johnson, focused attention on the appreciation of nature and wild places.

What government agencies and industry groups did not expect was the rapid growth and popularity of the environmental movement. For the first time, industry interests faced real opposition from both new and established groups who raised concerns about the damage caused by overgrazing. In 1964, the Classification and Multiple Use Act (CMU) required the BLM to apply the same standards for multiple use to rangelands as were required for national forests. Under Interior Secretary Stewart Udall, the federal government took a more proactive role in managing grazing, rather than simply responding to the demands of livestock owners.

A typical example of the shift in both governmental and public attitudes is Sequoia National Park. As described previously, massive increases in the number of visitors and vehicles had severely degraded the park's ecosystem. It was not until 1974 that a planning report underscored the warnings of infrastructure impacts and recommendations written in 1930. The NPS eventually removed 282 buildings, 1 million square feet of asphalt, buried propane tanks, and miles of overhead power lines, and converted the old village market into an interpretive center. Commercial enterprises were moved out of the park, the grove was restored, and a shuttle system plan is under development (Leonard 2005). These changes in park management would not have occurred were it not for the role of the environmental movement, especially in states like California.

Environmental groups have found a niche in the West; not only have the mainstream groups maintained a presence in cities like San Francisco, Missoula, and Santa Fe, but, in addition, hundreds of grassroots groups have mushroomed to deal with more local and regionalized concerns. The Federation of Fly Fishers (Livingston, Montana), High Country Citizens' Alliance (Crested Butte, Colorado), Nevada Tahoe Conservation District (Stateline, Nevada), and Center for Biological Diversity (Tucson, Arizona) are typical of the kinds of nongovernmental organizations that exist today as a result of the expansion of environmentalism in the West.

Some communities that were once dependent upon natural resource–based employers found a new base for their economy— tourism. Those near recreational areas turned timber jobs into service industry employment. In Flagstaff, Arizona, for instance, new federal regulations led to reductions in logging in the Coconino, Kaibab, and Prescott national forests, and the subsequent closing of local mills. But the city is less than two hours' drive from the Grand Canyon, and less than one hour from a ski area and the red rock area of Sedona, Arizona. Realizing that there was little to be gained by trying to stop a national change in natural resource policies, the city began to move toward a hospitality-based economy, building hundreds of new motel and hotel rooms, restaurants, and second homes for owners from Phoenix, Las Vegas, and California.

Not everyone rushed to become part of the tourist boom, though, and even environmental groups balked at large new developments and traffic leading into the Grand Canyon. Many were discouraged by the new Indian casinos being built on the interstate highways, and the mushrooming gasoline stations, mini markets, and chain restaurants that seemed to appear overnight. Others were dismayed by efforts to thin forests under the provisions of the Healthy Forests Restoration Act of 2003, concerned that the timber industry was trying to make a comeback under the guise of ecological restoration and wildfire prevention.

Another factor that has been advanced for the controversies that took place after 1960 is the redefinition of the concept of conservation. Gifford Pinchot's goal of national efficiency through scientific management, which had been the cornerstone of national policy throughout the first half of the twentieth century, was gradually replaced after World War II with a belief that commodity production should be secondary, rather than primary, to interests relating to outdoor recreation, wildlife protection, and preservation of scenic areas for aesthetic reasons. Pinchot's definition of multiple use was replaced by a new one—with new adversaries and supporters.

While interest in natural resource problems has waxed and waned since 1960 as some environmental issues have taken precedence over others, there has been one key shift in direction. In the late 1950s through the early 1970s, much of the public and governmental concern was over pollution, primarily water and air quality, and later, over hazardous waste and toxic pollution. A deadly smog siege in New York in 1962 resulting in eighty deaths focused attention on air quality in urban areas; the 1969 Santa Barbara oil spill represented a crisis in marine pollution. It was during this time that most of the major environmental laws were enacted by Congress, from the 1961 Federal Water Pollution Control Act to the Clean Air Act of 1963. By the time that the first Earth Day was held, on April 22, 1970, the public had been awakened to the environmental issues brought on by both technological change and by crisis. The National Environmental Policy Act of 1970 (NEPA), considered by many to be the foundation of modern environmental law, put new restrictions on government projects, requiring an assessment to determine what impact, if any, there might be on the overall environment. At the same time Congress began fine-tuning the earlier pollution laws, passing the Clean Air Act of 1970 and the Clean Water Act of 1972. By the mid-1970s energy issues, especially those related to supply and demand, were at the top of the political agenda. The United States, faced with the Arab oil embargo and energy crisis of 1973–1974, took a closer look at how regulations might affect domestic oil and gas production. From that time on, most pollution laws reflected incremental changes rather than wholesale reform.

By the 1980s and 1990s, pollution gave way to other concerns, many of which reflected a return to issues relating to natural resources. One legislative response that attempted to remedy the divergent objectives was the Multiple Use and Sustained Yield Act of 1960 (MUSYA). It was among the first of several attempts to balance the need for timber against the desire of citizens to use the national forests for outdoor recreation. Four years later, the 1964 Wilderness Act gave Congress, rather than the Forest Service, the authority to designate tracts of land as wilderness. The designation trumped the MUSYA, essentially eliminating any development that was permitted under the definition of multiple use. Despite the two statutes, timber harvesting continued as the dominant use of federal lands until the 1970s. With passage of the National Forest Management Act of 1976 (NFMA) and the Federal Land Policy and Management Act of 1976 (FLPMA), Congress made clear its intent to retain control of public lands while directing both the Forest Service and BLM to operate under the evolving principle of multiple use.

The center point of the land management debate became Alaska, which was still being homesteaded until the mid-1970s. In 1980, the Alaska National Interest Lands Conservation Act (ANILCA) placed 104 million acres of land under federal protection, much as forest reserves had been created in the late 1800s. Some of the region had already gained protection with the creation of the Arctic National Wildlife Refuge (ANWR), but some areas were set aside for future development because of the potential for oil and natural gas drilling and production. Environmental groups sought wilderness designation, but energy interests (and the Alaska congressional delegation) wanted the land to be leased to private companies for exploration. The conflict has not yet been resolved, despite more than twenty-five years of hearings, protests, and lobbying by both sides, most recently in 2006.

Public land issues resurfaced in the 1970s and into the new century because of several factors. The economic recession of the early 1970s resulted in outmigration from the frostbelt states because of industrial and manufacturing plant closures and rising unemployment. At the same time, sunbelt states experienced an influx of population, lower tax rates, and employment opportunities. Western states were criticized for taking advantage of the recession by increasing energy development, even though the temporary boom came with a high environmental cost. States' rights issues resurfaced over taxes on nonrenewable resources such a minerals, and some states viewed Eastern interests as interventionists, trying to impose their regulations on the still-frontier mentality of the West. Regionalism became pervasive.

Jimmy Carter failed to win any Western state electoral votes in the 1976 presidential election, and many leaders felt that the former Georgia governor had little knowledge of, or concern for, Western issues. Although he had benefited from the support of the environmental movement, by the end of 1979 one author proclaimed that "the long honeymoon of the environmental movement with Jimmy Carter is over" (Cawley 1993, 86). The election of Ronald Reagan in 1980 encouraged a recurrence of the Sagebrush Rebellions that had characterized several earlier conflicts over grazing rights on public lands. Wise use groups, such as the League for the Advancement of States' Rights (LASER) and the Center for the Defense of Free Enterprise, jumped on the Reagan bandwagon and tried to steer it in their direction. They expected that Reagan, who had served two terms as the governor of California, would understand the needs of Western ranchers, miners, and loggers, and that he would end the regulatory nuisances pushed forward by environmental groups during the 1960s and 1970s. They cheered when both Reagan and his secretary of Interior, James Watt, publicly supported wise use and condemned government intrusion.

Public rangelands re-emerged as the scene of conflict not only because of the Sagebrush Rebellion but also because of several major federal policies. With rangelands constituting about 70 percent of the Western states, grazing issues are a natural source of conflict. Critics have often complained that the BLM has failed in its role as a steward of natural resources, failing to inventory and monitor rangeland conditions, ignoring damage to riparian areas, and allowing uncontrolled spread of noxious and invasive species such as cheatgrass and star thistle. Environmental groups focused their complaints not only on the BLM but also on livestock owners who ignored the terms of federal grazing permits, and an industry that seemed to control the permit process and the BLM as well.

Cattle ranchers, in particular, have accused environmental organizations of trying to "lock up the land" from livestock production, putting family ranchers out of business and destroying Western culture at the same time. This topic, discussed further in Chapter 2, is centered in the West but has far-reaching policy and political implications reaching to Washington, D.C., and the growing influence of members of Congress representing Western states.

A benchmark event took place in 1990 with the designation of the Northern spotted owl as a threatened species. For years environmental groups had sought to have the owl listed under the 1973 Endangered Species Act because its natural habitat is primarily old growth forests in the Pacific Northwest. Timber interests argued that setting aside critical habitat for the bird would limit logging and, therefore, cause economic harm to local communities. Although the debate over forest management and practices like clear-cutting had been going on for years, the passage of the Endangered Species Act and the NFMA gave environmental organizations a legal foundation for changing the way forests were protected. Most of the attention has been focused on the owl, but by the early 1990s, 162 species in the region had been candidates for federal listing as threatened or endangered (Dombeck, Wood, and Williams 2003). The designation of each species brought new disputes, new protests, and new lawsuits.

Another major conflict involved timber in roadless areas. As early as 1952, the chief of the Forest Service had told Congress that unless funds were appropriated to build logging roads in the national forests, much of the most productive timber would be inaccessible. Most roads at the time were built by the timber companies, which assumed that they controlled access to them and the land they covered. After the Forest Service got into the road-building business using taxpayer funds, conflicts developed over not only ownership but also access. In 1962 the attorney general ruled that there must be reciprocal access, and the number of miles of roads skyrocketed.

As will be discussed further in Chapter 2, the 2001 Roadless Area Conservation Rule became the most controversial public issue ever faced by the Forest Service, and it remains unresolved even now. Hundreds of public meetings, more than 1.6 million public comments on the proposed rule, countless appeals and lawsuits, and angry protests by both supporters and opponents of roadless areas have led to flip-flops in policy for years. Nearly 60 million acres are considered roadless areas, the majority of which are not productive timber acreage. But millions are considered recreational areas that should be available for snowmobiles, off-road vehicles, and other motorized vehicles.

By the start of the twenty-first century, some Americans believed that the major environmental problems, such as air and water pollution, had been solved and were no longer of critical importance. Federal budget cuts to resource agencies had led to closures of popular campgrounds, fewer interpretative programs, less research and monitoring of resources, and less enforcement of existing laws. Under President George W. Bush, new regulations reduced public participation in natural resource decision-making, and while cooperative collaboration became the buzzword out of Washington, environmental groups criticized the president for rollbacks of protective legislation. By 2006 the president's political popularity had sunk to record lows, and many observers believed that the Republican majority in Congress would be soundly defeated in the midterm elections. Some even predicted that a Democrat would be elected president in 2008.

But those predictions had little to do with the environment. Despite an enduring presence in U.S. politics, the environmental movement had become fragmented and lacked visible leadership. Natural resource issues such as grazing on public land, logging in national forests, and mining royalties were not even on the public's radar screen. In most public opinion polls during the century's first decade, the environment rarely made the list of problems considered to be the most important. War, the economy, health care, and education headed most lists, leading two activists to question whether environmentalism was dead.

The chapter that follows does not attempt to answer that question, but it does outline the major natural resource conflicts that emerged from the twentieth century. Those issues remain on the political agenda, and with them have come new ways of settling disputes, negotiating conflict, and managing natural resources in the United States. Chapter 3 goes a step further, by exploring parallel resource issues on the global level. There are overlaps in some areas, but the level of conflict and the efforts by individuals and groups to deal with problems are often amazingly quite similar.

## References

Adams, David A. 1993. *Renewable Resource Policy: The Legal-Institutional Foundations*. Washington, DC: Island.

Cawley, R. McGreggor. 1993. *Federal Land, Western Anger: The Sagebrush Rebellion and Environmental Politics*. Lawrence: University Press of Kansas.

Clawson, Marion. 1983. *The Federal Lands Revisited.* Baltimore, MD: Johns Hopkins University Press.

Davis, Charles, ed. 2001. *Western Public Lands and Environmental Politics.* 2d ed. Boulder, CO: Westview.

Dombeck, Michael P., Christopher A. Wood, and Jack E. Williams. 2003. *From Conquest to Conservation: Our Public Lands Legacy.* Washington, DC: Island.

Kline, Benjamin. 2000. *First along the River: A Brief History of the U.S. Environmental Movement.* San Francisco: Acada.

Leonard, Bruce. 2005. "Returning the Land to the Giants." *National Parks* 79, no. 1 (winter): 18–22.

Miller, Char. 2001. *Gifford Pinchot and the Making of Modern Environmentalism.* Washington, DC: Island.

O'Sullivan, John L. 1839. "The Great Nation of Futurity." *United States Democratic Review* 6, no. 23: 426–430.

Schulman, Bruce J. 2005. "Governing Nature, Nurturing Government: Resource Management and the Development of the American States, 1900–1912." *Journal of Policy History* 17, no. 4: 375–403.

# 2

# Problems, Controversies, and Solutions

As outlined in Chapter 1, conflicts over natural resources began when the first settlers arrived on the coast of colonial America. Stories of potential wealth brought immigrants looking for land, timber, minerals, and adventure. They quickly began exploiting what they found, cutting down forests for ships, housing, and fuel, and clearing land for agriculture. Settlement meant claims and sales of vast parcels of land to private owners, who then broke the parcels up and sold them to others.

Unlike the colonial experience, much of the West was purchased by the newly emerging federal government. Many of the contemporary conflicts relating to natural resources in the United States have their roots in the West, in large part because that is where many of those resources are found. Much of the acreage that is federally owned and managed as national forests, rangelands, and protected areas is in Western states, where privately owned land is less common than in the East or South. Although there is a significant amount of mining in the East and Appalachia (primarily coal production), and significant amounts of petroleum and gas are extracted in the Southwest, the controversies that accompany resources in those regions are primarily related to the impact of fossil-fueled power plants on air quality and acid rain. Similarly, national forests have been designated in nearly every state. But in the South, timber sales have declined over the past fifty years, and the amount of lumber being harvested is not nearly at the scale of what takes place in the Pacific Northwest.

31

Livestock grazing on public land is a Western enterprise, and many of the nation's most visited parks and monuments are in the thirteen Western states.

This chapter looks at four of the nation's most enduring controversies involving minerals, oil, and natural gas exploration and drilling; policies relating to protected areas such as national parks and monuments; rangeland management, and livestock grazing on public lands; and timber and forest conflicts, usually over old growth trees. Although some of the conflicts are more localized than others, there is a pattern to many of the struggles being fought by environmental organizations and advocates, industry interests, local, state, and national governments, consumers, public interest groups, and regulatory agencies. Despite the intensity of the dispute, or the bitterness of the debate, there are efforts being made to create collaborative partnerships, to reach consensus on even the most intractable issues, and to devise new ways of managing the nation's natural resources. The four sections that follow provide an overview of the controversy, identify the primary stakeholders, and explore the strategies used, both successfully and unsuccessfully, to resolve conflict.

# Minerals, Oil, and Natural Gas Exploration and Drilling

Although the definition of *natural resource,* explained in Chapter 1, can be confusing, that is not the case with nonrenewable resources. They exist on earth in fixed amounts and generally cannot be replenished. A nonrenewable resource is considered depleted when 80 percent of its total estimated supply has been removed and used. Supplies of the remaining 20 percent are usually too expensive to find, extract, and process. Only a few nonrenewables, including aluminum, can be used and then recycled; most are finite.

Minerals, the first class of nonrenewable resources, consist of chemical elements or compounds, usually in solid form, that occur naturally in the earth's crust, the lithosphere. Minerals were produced through geochemical processes that occurred over hundreds of millions of years. They generally fall into two categories: metallic, such as iron, aluminum, copper, and tin, and nonmetallic, such as sand and salt. Another class of natural resources related to mining—fossil fuels—includes coal, natural gas, and petroleum, or oil. Unlike minerals, fossil fuels are the remains of ancient plants and animals that have been compressed for millions of years by geophysical processes (Cable and Cable 1995).

The U.S. economy has been driven, in large part, by natural resources discovered from the beginning of western expansion in the 1800s and into the twenty-first century. In California, the discovery of gold in 1849 led to a rush of settlement and exploration, even though the land belonged to Mexico and was occupied by U.S. troops. Another gold rush in the Klondike in 1896 made the purchase of Alaska an important addition to the territories of the United States. In Wyoming, the discovery of one of the world's largest oilfields in the early 1900s caused cycles of boom and bust that continued until a second period of expansive growth in 2000. King Coal dominated U.S. fossil fuel policy for decades until other forms of energy were discovered and produced.

Similarly, mining regulations and laws have evolved gradually, from the mid-1800s when informal mining districts and codes were created, to the enactment of the first federal law in 1866, to the 1872 Mining Law, which continues to govern much of the industry today. For the most part, the mining, oil, and natural gas industries have been regulated by economic liberalism, the belief that government and society should have minimal interaction, and then only as a means of facilitating additional energy production (Klyza 1996).

The 1872 legislation was designed to bring structure, even if minimal in its impact, to the growing mining industry. The land remains under federal ownership, with the leases generating revenue, estimated at about $2 billion per year in royalties, split evenly between the state and the federal government. The Mineral Leasing Act of 1920 gave 37.5 percent of mineral leasing revenues to the states, with another 52.5 percent allocated to the federal Reclamation Fund for water projects.

## Supply and Demand

Understanding the role of minerals and fossil fuels is critical because of the realities of supply and demand, and because those supplies are dwindling in the face of increasing demand. Federal agencies, including the U.S. Geological Survey (USGS), have warned about the domestic reserves of minerals, which are considered adequate for the next several decades. The exceptions are chromium, cobalt, platinum, tin, gold, and palladium. But the USGS says that present reserves of other key minerals will not satisfy U.S. demand for more than 100 years without increased recycling, conservation, or substitution by other minerals. Identified coal reserves in the United States are expected to last another 300 years (Cable and Cable 1995).

The status of natural gas and oil are quite different. A half-century ago, a Shell Oil geophysicist, M. King Hubbert, proposed that there would come a time when the amount of oil that the United States could pump out of the ground would reach its peak and then begin to fall. That peak occurred in 1970, and then supplies dwindled while demand continued to rise. The United States began to import oil from other countries, increasingly those in the Middle East; by 2006, more than 60 percent of the country's oil supplies were coming from abroad. Worse yet, producers believe that within the next fifteen years, the world's oil supply will also reach its peak (Jenkins 2005). Even if more oil can be found, the resource still remains finite, and conflicts over supply are inevitable.

When U.S. gasoline prices jumped past the $3 per gallon mark in 2005, consumers were angered by reports that oil companies were experiencing record profits. Industry officials pointed to the ups and downs of oil prices: during the 1980s and 1990s, the price of a barrel of oil ranged from nearly $100 per barrel down to about $10 per barrel. The dry spells when many companies declare bankruptcy can sometimes be offset by periods of growth and consumer demand, they note, and also affect other segments of the economy. In 1980, Congress imposed a windfall profits tax when the war between Iraq and Iran caused U.S. prices to skyrocket. The tax was withdrawn in 1988, after countries such as Saudi Arabia expanded their operations and U.S. profits fell dramatically. Some analysts believe that imposition of another windfall profits tax now, even with the price of oil approaching $80 a barrel, would put the United States at a competitive disadvantage in the international market. At the same time it is unlikely that the market for oil and gas will be reduced, because of increased demand in developing countries such as China.

There are two major conflicts that typify contemporary problems relating to these forms of nonrenewable natural resources: the calculation of value and a fair return for extraction of resources, and environmental and infrastructure damage caused by industries and processes.

## Extraction and Money

The issues of value and return are not ones that can be dealt with simply by consulting accountants and attorneys. How much should companies pay to the government, and, indirectly, to citizens, for the rights to conduct mining or drilling operations on publicly owned land? That question has frustrated both industry interests and government officials since the 1800s. The debate has broadened to include not just the right to conduct operations within the public domain but also the question of reducing and mitigating the damage these industries cause to the environment.

To help deal with the value issue, several states have enacted severance taxes imposed on extractive industries like mining, oil, and natural gas. The taxes are based on the premise that there is a value attached to the minerals severed from the earth, and that the industry that receives the benefits (profits) ought to pay back a portion of those revenues to the state. In the past, severance taxes have been used for funding schools and government services such as fighting wildfires and road building, producing budget surpluses and incentives for states to create trust funds for projects such as college scholarships, economic assistance, preserving wildlife habitat, and expanding health care insurance. New Mexico received about $1.8 billion in royalties and taxes in 2005, based on nearly $12 billion worth of oil and gas production; Colorado's severance tax rate of about 2 percent on the $8 billion in 2005 production resulted in considerably less. From 2000 to 2004 annual severance tax revenues varied, from a low of $23 million to a high of $107 million, making financial planning problematic (Ring 2005).

In November 2005, another attempt at mining reform was made when the House of Representatives approved an obscure provision in the budget bill that would increase the price that mining companies pay, currently $2.50–$5 an acre, to $1,000 an acre or fair market value, whichever is higher. According to the Congressional Budget Office, the federal government could realize $158 million over the next five years, ostensibly to increase revenues that would then be applied to the federal deficit (Burkhart 2005). While this might seem like a way of resolving the conflict over the low land prices now in effect, the impact would have a spillover effect on mineral rights. Under the budget proposal, the land sales would be accompanied by the rights to any minerals found on the land, ending the existing royalty system.

## Damage from Mining Activities

There are many types of damage attributable to these industries and practices, ranging from roads that bear the weight of heavy equipment to piles of tailings from mines, many of which are abandoned. The burden of infrastructure damage falls primarily on local governments. The environmental impacts of mining and oil and natural gas drilling affect a much larger population and are long term. Water quality is often compromised as runoff from mining (acids, silt, and toxic substances) seeps into the ground and percolates down into groundwater, or finds its way into surface waters. Pollution from mining activities, such as dust and other particulate matter, is known to cause damage to public health and affect air quality. Waste, including piles of leftover rock and materials, must be hauled to another site, buried, or otherwise recovered.

The history of dealing with the environmental impact of mining and fossil-fuel production in the United States is not pretty. Nineteenth-century mining operations were frequently described in graphic terms. One writer who visited the lead mines southwest of St. Louis in 1818 told of winding along a countryside of "pits, heaps of gravel, and spars, and other rubbish constantly accumulating." A gold mine operation near Dahlonega, Georgia, was observed as having "[a]ll the valleys being dug up and washed gravel thrown into heaps, their beauty was entirely destroyed, and the scene resembled a series of brickyards." And Jamestown, North Carolina, was pictured as "turned topsy-turvy by the gold diggers, who had utterly ruined the valley for agricultural purposes" (Smith 1987).

One strategy for dealing with the environmental damage caused by mining has been legislation, although it often comes long after a problem has been identified. In 1977, Congress addressed the issue of coal mining and reclamation with passage of the Surface Mining Control and Reclamation Act, placing responsibility for environmental cleanup under the Office of Surface Mining (OSM) within the U.S. Department of the Interior. The law requires active coal mine operators to reclaim the land as they go; previously, mines were simply abandoned, creating hazards from open portals and vertical openings, piles and embankments, and clifflike high walls. The OSM has a three-region organizational structure that handles mines in twenty-five states and four Indian tribes in the West, midcontinent, and Appalachia. Environmental cleanup costs are paid by a combination of state and federal funds, and by tonnage-based reclamation fees paid by active coal producers. This system has reduced the types of conflicts common to the cleanup of hazardous waste sites under the nation's Superfund legislation.

## Oil and the Arctic National Wildlife Refuge

The best known, and still unresolved, issue that typifies the environmental conflicts over oil exploration and drilling in the United States is that of the Arctic National Wildlife Refuge (ANWR). The value of Alaskan wilderness was first publicized by Bob Marshall, one of the foremost environmental activists of the 1930s. Marshall sought to have the region, which was not yet a state, protected in pristine condition in perpetuity. The territory was seen as valuable for other reasons, however. The Department of Defense saw the potential for military bases, especially during the Cold War with the Soviet Union. Whalers and fur traders, who had long used Alaska for fishing, hunting, and trapping, assumed that the region would stay much the same as it had always been, remote and forsaken by commercial development. Real estate investors saw the potential for an economic boom if the issue of transportation could be settled. By 1959, when Alaska became a state, political and economic interests were paying more attention to the area than were environmental organizations, which were only beginning to recognize the environmental potential and possibilities for degradation.

During the 1960s, Alaska became more than just a far-off extension of the continental United States. The Atlantic Richfield Oil Company had discovered oil in the Prudhoe Bay area in 1967, and development of new energy sources was frenzied. President Richard Nixon chose Walter Hickel, the state's governor, as his secretary of the Department of the Interior, giving him the inside track on any and all policies affecting Alaska. Hickel had been in a lengthy pursuit of an economic goal—the building of a pipeline to carry oil from the state's North Slope to the port of Valdez, where it could be loaded onto tankers for shipping.

Hickel's plans coincided with the growing influence of the environmental movement, whose leaders were looking for a focusing event that would push their activist agenda from the drawing board to Washington, D.C. That event was the January 29, 1969, Santa Barbara oil spill; a Union Oil drilling rig six miles off the coast ruptured, causing 200,000 gallons of crude oil to bubble to the surface. The resulting 800-square-mile oil slick spread over 35 miles of the California coastline and south to the islands nearby. Workers spent nearly two weeks trying to control the leaking oil well, which caused the deaths of thousands of shorebirds and marine mammals.

Three factors relating to the oil spill galvanized the grassroots movement that subsequently developed. First, the news media coverage was unsurpassed compared with any previous environmental event. The imagery was impossible to ignore, as dead seals, dolphins, and birds covered in oil washed ashore. Second, subsequent investigations revealed that the USGS had given the company permission to cut corners and operate the drilling platform with materials below federal and state standards. Because the site was beyond California's three-mile coastal zone, the company did not have to comply with state standards; the backlash against Union Oil was immediate. Third, there were no partisan forces at work. Neither Republicans nor Democrats could support Union Oil's operations, given the level of damage and the public uproar that followed. The event virtually ruined the oil industry's reputation, and new outraged activists cut up their oil company credit cards, boycotted gas stations, and called upon Congress to strengthen regulations and laws on offshore drilling.

Less than two weeks later, on February 10, 1969, plans were announced for the Trans-Alaska Pipeline System (TAPS), a proposed 798-mile-long conduit for oil that would vastly increase U.S. energy capabilities. It is unclear whether Hickel and the oil companies underestimated the impact of the Santa Barbara oil spill, or the strength of the environmental organizations that had seized upon the event to mobilize support for stronger legislation. Three major lawsuits against the project were filed by opponents, and public skepticism over claims that there would be minimal environmental damage grew. By early 1971, President Nixon had replaced Hickel, and a new Department of the Interior secretary, Rogers Morton, held hearings on the pipeline. Environmentalists formed a new group, the Alaska Public Interest Coalition, to coordinate opposition to the project, and the oil industry remained steadfast in its position that new sources of oil were essential to the nation's economy.

Even after the department released documents in 1972 that favored oil drilling to reduce the reliance on foreign sources, environmentalists continued to litigate against the agency, using the new National Environmental Policy Act (NEPA) and the Mineral Leasing Act. Public sentiment against TAPS was tempered, however, by the 1973 Arab oil embargo, when the images of birds stuck in tar were replaced by scenes of hundreds of cars stretched in lines waiting to fill up on gasoline. By the end of the year, the pipeline project had passed through Congress and was signed by the president.

Having lost one battle to preserve Alaskan wilderness, environmental organizations turned to existing legislation rather than litigation. In 1960, the federal government had set aside 8.9 million acres as the Arctic National Wildlife Range; another 3.7 million acres were added later under the Alaska Native Claims Settlement Act (ANCSA) of 1971. In 1980, the Alaska National Interest Lands Conservation Act (ANILCA) became another protective measure supported by environmental groups, reclassifying the initial 8.9 million acres as the Arctic National Wildlife Refuge and adding 9.1 million acres of adjoining public lands. The pipeline corridor is adjacent to the ANWR, an area almost as large as New England (Layser 2006).

The current conflict is almost a repeat of what happened with TAPS. A consortium of environmental organizations is working with the Canadian government and indigenous groups reliant upon the caribou herds that traverse the refuge to keep the area pristine. Oil interests point to provisions of ANILCA that exclude from protection areas with significant development potential. Other stakeholders include the Inupiat Eskimos, who support further exploration because oil drilling has brought prosperity to their people, and the Coalition for American Energy Security, which includes the Inupiat and the motorized recreational vehicle industry and users, transportation companies that would profit from shipping, labor unions, and agricultural interests.

Supporters of exploration and drilling cite the need for developing domestic sources of fuel and the crisis in the Persian Gulf as prime reasons for reducing U.S. dependence upon foreign oil. The Coastal Plain would become the third largest oil field in the country, according to the Department of the Interior. The war in Iraq helped bolster their position by framing ANWR as part of the debate over national security. Opponents have relied less on the issues of damage to wildlife and environmental degradation, and have argued that the answer to oil dependency lies with conservation of fossil fuels and reduced energy use, not more drilling. The two sides dispute the potential amount of fuel within ANWR, the number of jobs that could be created, and projections on the price of oil.

But the images of an environmental disaster that worked so well after the Santa Barbara oil spill have not resonated as well with ANWR. Since 1986, Congress has gone back and forth on the future of oil drilling in the refuge. Members of Alaska's congressional delegation, who have usually held positions of power on natural resource committees, have sometimes lost out to other committee leaders who also claim jurisdiction over the issue. Focusing events, such as the 11 million gallons of oil spilled along Alaska's Prince William Sound by the *Exxon Valdez* in 1989, have set supporters of drilling back, but only temporarily. The doom-and-gloom predictions by environmentalists that the region had suffered catastrophic damage never came to pass. The 1990 Persian Gulf War against Saddam Hussein after his country's invasion of Kuwait again brought forth calls for congressional action on ANWR for reasons of national security. Leasing proposals were damaged by allegations of corruption by the owners of the TAPS and a world oil glut when Saudi Arabia increased its production.

Members of Congress have periodically reopened the ANWR debate, with varying degrees of success. Supporters looked to President George W. Bush and his promise of energy independence, and oil companies pledged to use drilling operations that would mitigate any environmental damage. The federal budget was held hostage to provisions relating to ANWR, although the political climate seemed to favor drilling as gas prices rose again. What this says about conflicts related to oil exploration and drilling is that there seems to be no middle ground. Environmental organizations have vowed to tie up proposals in court for years, arguing that no drilling is the only acceptable outcome. Oil companies keep relying on rising gas prices and issues relating to national security and the Middle East. There is, it seems, no room for compromise or collaboration, only periodic forays into enemy territory.

One of the newer controversies will be dependent upon where the next major phase of exploration and drilling takes place. Although the debate over ANWR is certainly not over, supporters of the oil drilling project are using a different strategy, which is likely to prolong the conflict. Rather than relying upon Alaskan supplies, energy producers are shifting to new sources in different places. One analyst has predicted that 400,000 new natural gas wells will be built in the Rocky Mountain region over the next fifteen years. A recent report by the RAND Corporation estimates that enough oil can be produced from oil shale in Wyoming, Colorado, and Utah to more than triple Saudi Arabia's proven oil reserves. Oil shale is actually a hard rock called marlstone that, when heated, produces a low-grade oil. The technology is still in the early stages, and there are major environmental consequences to digging mines, using large amounts of water, dealing with leftover shale, and refining and transporting the oil (Lay 2005).

Commercial production is probably a long way off, even though Congress and the Department of the Interior are supportive of oil shale research and the Bureau of Land Management has begun a ten-year program to lease small parcels of public land to try out new methods of production. But as a political strategy, focusing attention on other sources of fossil fuel, whether natural gas or oil shale, is a smart way to keep environmental organizations busy. If drilling in ANWR continues to be shuffled on and off the political agenda, energy companies can keep the heat on other projects, forcing opponents to spend more resources and time on less well-known controversies. The conflicts never really end, they just get moved to other states.

# Protected Areas Policy

In 1916, the British ambassador to the United States, James Bryce, called the creation of the national park system "the best idea America ever had." His comments reflect what many people believe to be one of the major accomplishments in natural resource policy—the development of a system to protect the country's natural and scenic areas. But the National Park Service (NPS) and the lands and sites that the agency administers have not been without controversy.

The first designated protected area in the United States, Yellowstone National Park, established in 1872, resulted from the romanticized view of the Western frontier that had taken hold after a decade of exploration and discovery. But eighteen years passed before a second park was designated, and many members of Congress were not supportive of the concept of withdrawing lands from public use, or of the cost of protecting and maintaining new parks. Entrance fees were instituted at Mount Rainier in 1908, Crater Lake in 1911, and Yosemite National Park in 1913.

Each area was managed individually under the leadership of three different cabinet agencies, dependent upon where they were located and the purpose for their preservation. Three departments—War, Interior, and Agriculture—held jurisdiction over the parks; there was no "system" to coordinate park management or policies. With passage of the 1916 Organic Act, park officials were given the seemingly dichotomous roles of preserving nature and providing recreational opportunities, placed under the new National Park Service in the Department of the Interior (Dilsaver 1994, 2).

The management style of the NPS began to change during the 1960s and 1970s. From the passage of the Organic Act on, the primary purpose of the administrative structure was the preservation of scenery, and later, sites of historical or cultural interest. Decisions on how best to do that were mostly routine and based upon how much money was available from one fiscal year's budget to the next. But with the advent of the environmental movement and the accompanying advances in the field of ecology, science began to play a much more important role in decision-making. External reports and analyses, many of them written by environmental organizations, focused on the need to base management on scientific research and ecological concerns, rather than on visitation, legislative pressures, or operational needs. Many outside observers were critical of the lack of information and monitoring of species and the lack of any systematic planning or inventories. While some efforts to expand staffing and employees' natural resource expertise were made in the 1980s, by the end of the decade other needs had been prioritized (Sellars 1997).

During the 1990s, the federal emphasis on efficiency and the reinventing government movement also affected protected areas such as national parks. Bureaucratic agencies responsible for the delivery of public goods and services, including the NPS, were subjected to review and audits as part of the performance management focus of President Bill Clinton and Vice President Al Gore, Jr. As a result, park managers were increasingly forced to function with fewer resources, leading to a greater emphasis on resort-type recreation and less emphasis on protection of natural resources (Lowry 2001).

Although the issues surrounding protected areas are numerous, there are three that have created the most problems and controversy, all of which are related to funding: the fee demonstration program enacted by Congress in 1996, the lack of sufficient staff and creation of a maintenance backlog within the parks' infrastructure, and the increasing number of protected area designations.

## Fee Demonstration Program

All four of the nation's land management agencies have been creative in finding ways to pay for basic operational expenses, as well as programmatic and research needs, especially during an era when parks, monuments, and other protected areas are facing budget cuts. Most of the park units charge a daily or weekly entrance fee, and some services within the units are also fee-based. Other forms of protected areas, such as battlefields, rivers, trails, and urban parks, where there is unfettered access, generally do not charge a fee.

In 1996, Congress enacted the Recreational Fee Demonstration Program (RFDP) to help pay for the increasing number of visitors and rising operational costs at protected areas managed by the NPS, U.S. Forest Service, Bureau of Land Management, and the Fish and Wildlife Service. The fee demo program, as it is known, was established on a trial basis in November 1996 at forty-seven NPS units. Fees were increased in some units, and collected for the first time in others. The plan was to return 80 percent of the new revenues to the sites where they were collected. The other 20 percent of the additional revenue was to be distributed by the director of the NPS to special emphasis areas in the parks. In March 1997, forty-five NPS units were added to the program, with another seven added in April 1998. The legislation called for a trial period of three years, and the program was extended five times.

Additional fees were implemented in 1997 by the Forest Service: the Adventure Pass, which is charged for parking a vehicle on any road for any amount of time within the boundaries of four national forests in southern California; and the Northwest Forest Pass, implemented in 2000 at twenty national forests in Oregon and Washington and in the North Cascades National Park Service Complex. That latter fee applies to parking and trail and facility use. Violations of any of the fee demo programs are infractions (less serious than a misdemeanor), resulting in a maximum $100 fine. Other fee-based programs include the America the Beautiful Pass, an interagency pass that covers the entrance fee and recreation amenity fee for federal recreational lands.

Agency officials estimate that since the fee demo program was implemented, about half of the revenues have gone to reduce the backlog of infrastructure maintenance. Trails have been approved and improved to comply with federal laws such as the Americans with Disabilities Act. Historic structures have been rehabilitated, cultural landscapes have been restored, and museum objects have been protected. Additional funds have been allocated for visitor orientation and education (mostly capital improvements such as amphitheater repair) and to replace campground structures and informational signs.

Between the first full year of the program's operation, 1997, and 2004, several groups and agencies have published reports reflecting the controversial nature of the experiment. The General Accounting Office (now called the Government Accountability Office, but still using the abbreviation GAO) reported in November 1998 that in fiscal year 1996, the last year before the demonstration program was implemented, the four agencies collected a total of about $93.3 million in fees from visitors (U.S. General Accounting Office 1998). In fiscal year 1997, the total collected at fee demonstration sites was $123.8 million of the $144.6 million in fees collected systemwide. For 1998, nearly 85 percent of the fee demonstration site revenues ($160 million) came from National Park Service sites, with 11 percent from Forest Service sites, 2.1 percent from the Fish and Wildlife Service, and 2 percent from the Bureau of Land Management.

An August 2002 report to Congress by the NPS and BLM, required by the initial fee demo legislation, found that the sites have received additional funds to undertake projects that might not have been possible otherwise, although the projects themselves could have been completed in a more timely manner. As of 2000, the sites surveyed had completed only about 11 percent of the fee demo projects (U.S. Department of the Interior 2002). The National Park Service commissioned a survey of public attitudes toward fees that, unlike some other studies, included a nationwide sampling of visitors and nonvisitors. The June 2003 report found that 95 percent of Americans are not familiar with the fee demo program at all; among the people who are familiar with the program, 94 percent support it. Eighty percent of NPS system visitors think that the fee they paid was "just about right" for the value received, while 11 percent said they had paid "too much." By a two to one margin, respondents favored lower entrance fees, with additional fees for services utilized, rather than one large, all-inclusive entrance fee. Perhaps the most important finding of all was that entrance fees do not present a significant barrier to visitation of NPS units (U.S. Department of the Interior 2003).

While there has been support for entrance fees to national parks, fee demo sites managed by the Forest Service have been the target of many complaints. Groups such as the Arizona No-Fee Coalition have told their supporters not to use national forests in which fees are charged, and if they do enter a fee demo site, not to buy a pass. The organization notes that many visitors refuse to purchase a pass, and often nothing happens as a result. The GAO has been relatively critical of the agency, questioning the lack of consistent information on the cost of collecting the fees and on where the revenues have been spent. Noting that the accuracy of some information was questionable, the GAO gave credibility to opponents who argue that no one knows where the money goes (U.S. General Accounting Office 2003).

A major programmatic change took place in December 2004, when President George W. Bush signed the omnibus appropriations bill containing a rider attached by Rep. Ralph Regula (R-OH) that changed the intent and management of the fee system. The legislation, the Federal Lands Recreation Enhancement Act, defined three levels of fees with a ten-year fee authorization. The new structure includes:

1. Standard Amenity Recreation Fee: applied to developed areas with at least six amenities: parking, a permanent toilet, permanent trash receptacle, interpretive sign, picnic tables, and security services;
2. Expanded Amenity Recreation Fee: applied to campgrounds, developed boat launches, developed swimming areas, and cabin or equipment rental;
3. Special Recreation Permit Fee: applied to commercial users and organized events. The legislation also clarified activities for which a fee could not be levied, including
   - Parking, undesignated parking, or picnicking along roads or trailsides;
   - General access;
   - Dispersed areas with low or no investment, unless specifically authorized by law;

- Persons who are driving through, walking through, boating through, horseback riding through, or hiking through federal recreational lands and waters without using the services;
- Camping at undeveloped sites that do not provide a minimum of six amenity services or facilities;
- Use of overlooks or scenic pullouts.

In what might seem like a major change, the law prohibited entrance fees for federal recreational lands and waters managed by the Bureau of Land Management, the Bureau of Reclamation (newly added to the program), or the Forest Service (the Fish and Wildlife Service is authorized to collect entrance fees). But there are numerous exceptions that allow the other three agencies to charge fees for a National Conservation Area, a National Volcanic Monument, a destination visitor or interpretive center, and other sites that provide significant opportunities for recreation, represent significant federal investment, or in which fees can be efficiently collected. Ironically, fees charged at sites managed by these agencies have been among the most controversial, even though the revenues collected are far less than those of the NPS.

Unlike some natural resource conflicts in which there are signs of collaboration, the fee demo program is an issue that sharply divides those who support the concept of having users pay to enter protected areas and those who believe that the "public" in public lands entitles them to free visitation and use. The Sierra Club has opposed the program from the very beginning because the organization believes that it raises serious questions about free public access and social equity, since the fees place a regressive tax burden on those with lower incomes and those who use the lands frequently. They also consider the fee demo program to be "double taxation [that] can lead to increased commercialization, privatization, and motorization of America's public lands" (Sierra Club 2006).

Although most mainstream groups have been measured in their public opposition, several smaller organizations have made the fee demo program the focus of bitter protests. Among the latter are the organizations Wild Wilderness, Keep the Sespe Wild Committee, the Western Slope No-Fee Coalition, and the group Free Our Forests. These groups are generally well connected to one another, with members who feel strongly about the issue when it affects sites they visit regularly.

The Western Slope No-Fee Coalition, based in Norwood, Colorado, is one of the more activist opposition groups, calling the fee demo program the Recreational Access Tax, or RAT. The coalition has accused land management agencies of circumventing agency guidelines by illegally charging fees for dispersed areas, general access, and recreational use of undeveloped land. Although the group says that they have attempted to work with congressional committee staff to develop reasonable changes, they have been unsuccessful and now are actively encouraging defiance (Western Slope No-Fee Coalition 2006).

One of the groups whose interests are affected by the program, but which also has other goals, is the American Hiking Society (AHS). This organization supports the recreational fee program in concept, noting that revenues could provide funding to land management agencies with severe budget shortages. Since some of the funding is meant to support trails and recreational resources and facilities, the group's members benefit directly. However, the AHS, in conjunction with the Outdoor Industry Association, has also urged Congress to conduct comprehensive oversight hearings, including testimony from representatives from affected constituency groups, before permanently authorizing the program. Their concerns, some of which have never been addressed, are similar to those raised by the organizations most opposed to the program. The difference is that their membership continues to function with other activities beyond the fee demo program (American Hiking Society 2006).

Several states, including Oregon, Colorado, and New Hampshire, have passed resolutions opposing the fee demo program, as have city governments and commercial interests. There are numerous state-level no-fee coalitions, and opposition from recreation organizations such as American Whitewater. Finding support for the fees outside the federal bureaucracy is much more difficult. The National Association of Gateway Communities, the American Recreation Coalition, and both the Southeast and Western States Tourism Policy Councils have rallied behind the federal government's position. Representatives have testified before congressional hearings, but their influence pales in comparison to that of the leadership of agencies within the executive branch.

Passage of the ten-year extension of the fee demo program does not mean that the conflict is over. Some opponents believe that the new fee structure, confusing and often contradictory implementation guidelines, and more negative reports about the program's operation will eventually reach congressional ears. After more than ten years of support, however, the federal government and the affected agencies seem more committed than ever to making the limited "pilot" program an extensive and permanent one.

## Staffing and Infrastructure Maintenance

Staffing is a very visible problem that affects one of the primary activities within protected areas—public outreach and interpretation. In 2005, Olympic National Park reduced the operating hours of its visitor centers during the summer for the first time in its history, and at Acadia National Park in Maine, a $500,000 reduction in the fiscal year budget led to a reduction in the number of the park's seasonal interpretative employees. California's Death Valley National Park has fifteen protection rangers to patrol 3.4 million acres of land, down from twenty-three a few years ago. Because of the need to protect visitors from temperature extremes and also to protect historical artifacts and petroglyphs, the staff has minimal resources for answering questions or providing educational programs in the park (National Parks Conservation Association 2005a.)

The National Parks Conservation Association (NPCA), a nonprofit organization, says that on average, national parks operate with only two-thirds of the money needed to cover annual operating expenses—or a shortfall of more than $600 million per year. Estimates as to how much it would cost to repair and replace the infrastructure within the national parks vary considerably. The NPCA says that there is a deferred maintenance backlog of between $4.1 and $6.8 billion, more than double the entire yearly budget for the NPS. A 1998 report produced by the GAO found that the NPS estimate of $6.1 billion in its maintenance backlog was inaccurate, being based on information that was often unreliable and an imprecise definition of the term *maintenance*. The figure included, for instance, construction of new employee housing, land acquisition, and other activities that would not generally be considered in the same category (U.S. General Accounting Office 1998).

One solution has been the creation of private organizations that have attempted to raise money and other forms of assistance for these sites, absent sufficient federal funding. In 1935 the National Park Trust was established as a mechanism for the private sector to donate land, gifts, and funds to the national parks; Congress established the National Park Foundation (NPF) in 1967 as the successor to the trust. Its first grants were made in 1968 for restoration projects at Theodore Roosevelt's birthplace and home in New York. The NPF has acquired land for the Blue Ridge Parkway, at Gettysburg, in the Virgin Islands, and in Death Valley, and has sponsored research, surveys, and studies, staged exhibitions, commissioned public opinion polls, published guidebooks, launched a children's environmental education curriculum, and provided executive-level volunteers. Among its major accomplishments has been the foundation's development of partnerships with private corporations, including American Airlines, Ford Motor Company, Kodak, Discovery Communications, and Unilever (National Park Foundation 2005).

Other private-sector efforts have focused on specific sites. In Grand Canyon National Park, for instance, where some of the first endangered California condors were released, there is only one wildlife biologist on staff, who is responsible for more than 1 million acres and more than 600 wildlife species. The Grand Canyon National Park Foundation, a nonprofit support group, has raised more than $65,000 since 2001 to help monitor the condors and provide public outreach about them. There are more than 700 miles of trails within the park, some of which are more than 100 years old; the shortfall for trail maintenance is estimated at $1 million per year. Volunteers from other nonprofits, such as the Grand Canyon Trust, are sometimes recruited to help control the spread of invasive plants, but their work is sporadic and insufficient to cover more than a small area.

One of the more controversial proposals has been to sell off the national parks, along with the naming rights to visitor centers and trails, partly to help balance the federal budget. Rep. Richard Pombo (R-CA), chair of the House Committee on Resources, floated the idea in September 2005, contending that the sales would help eliminate many of the inherent problems some parks face in trying to make ends meet. He later said that the idea was just a "conversation starter," and some critics noted that such extreme proposals were not serious attempts at solving the real issues faced by the National Park Service (Burkhart 2005).

But the NPS has moved forward on a new policy of aggressively seeking corporate sponsorship of park projects and facilities. Sponsors, including alcohol, tobacco, and gaming companies, could become the "Proud Partner" within a park, and use the NPS logo alongside their own. "This starts a slow commercialization of the national park system," says Jeff Ruch, executive director of Public Employees for Environmental Responsibility. "What will be allowed stops just short of licensing ads for 'The Official Beer of Yosemite' or 'Old Faithful, Brought to You by Viagra'" (Public Employees 2005).

In 2005, the NPS received only $17 million in private donations; Interior Secretary Gale Norton called the plan to seek out sponsors an "exciting new approach" for broadening the funding base for national parks. But critics argue that corporate sponsorship will mean that park managers will be under more pressure to appease donors. As Ruch notes, "Influence peddling will soon become a major recreational activity in our national parks."

## Too Many Protected Areas?

Funding problems are exacerbated by the growing number of areas that have been designated for federal protection. When the agency was created in 1916, the NPS was responsible for fourteen national parks, twenty-one national monuments, and two national reservations. The 1970 amendments to the 1916 National Park Service Organic Act created a wide-ranging list of categories, termed "units," greatly expanding the number of sites that are now under the jurisdiction of the NPS, including

- 1 International Historic Site
- 11 National Battlefields
- 3 National Battlefield Parks
- 1 National Battlefield Site
- 41 National Historical Parks
- 77 National Historical Sites
- 4 National Lakeshores
- 74 National Monuments
- 29 National Memorials
- 9 National Military Parks
- 57 National Parks
- 4 National Parkways
- 18 National Preserves
- 2 National Reserves
- 5 National Rivers
- 18 National Recreation Areas

- 10 National Seashores
- 3 National Scenic Trails
- 10 Wild and Scenic Rivers
- 11 Nondesignated Units

The NPS also is responsible for the nation's fourteen national cemeteries, all of which are administered in conjunction with an associated unit and are not accounted for separately (National Parks Conservation Association 2005b).

The process of setting aside protected areas in the United States is somewhat complicated and highly politicized. When a park is first proposed, Congress can enact legislation to create it, or members can ask the secretary of the Interior to conduct a study to determine whether there is sufficient funding for purchase or maintenance. The criteria for park designations are

- Is the site nationally significant? Is it an outstanding example of a particular type of resource? Does it possess exceptional value or quality in illustrating or interpreting the natural or cultural themes of our nation's heritage? Does it offer superlative opportunities for public enjoyment or scientific study? Does it retain a high degree of integrity as a true, accurate, and relatively unspoiled example of a resource?
- Is the site suitable, and not previously represented in the park system?
- Is the area manageable? Can it be protected? Is it capable of efficient administration by the Park Service at a reasonable cost?
- Is the NPS an appropriate manager? Or are there private entities that would be in a better position or are already managing the site successfully? (National Parks Conservation Association 2005a)

Many of the sites have been controversial because they were designated under political pressure rather than based on recommendations by organizations, or were designated with little public support. For example, an NPS study of a proposed site, the Moccasin Beds in Chattanooga, Tennessee, showed that the tract of land was a significant Cherokee Indian settlement. But there was already a golf course and another facility on the riverside land, so the NPS objected to including the site. A local member of

Congress objected to the agency's findings and secured enough support for the necessary legislation to be enacted. Presidents Jimmy Carter and Bill Clinton used their power under the 1906 Antiquities Act to designate parks and preserves in Alaska, as well as national monuments in the West. Under the administration of George W. Bush, the Office of Management and Budget advised the NPS to make a blanket recommendation against any new park units, including potential units clearly qualified and seriously threatened, according to the NPCA.

However, despite that policy, Congress designated the crash site of Flight 93 in Pennsylvania (where an aircraft was diverted during the September 11, 2001, terrorist attacks) as a national memorial, and the administration agreed to support designation of a small World War II historic site on the island of Palau, as well as a battlefield in Virginia's Shenandoah Valley as the 388th unit in the NPS system. Recommendations by the NPCA for inclusion of other sites, such as Puerto Rico's Bioluminescent Bay (home to a microscopic organism that glows bright green at night) and Iowa's Loess Hills (a unique geological formation), have not been accepted (ibid.).

Supporters of the more established park units, especially those with the most visitors, believe that the NPS has become a client of members of Congress who want to bring attention to a site, no matter how insignificant, in their district. Presumably, this would appease areas seeking economic development, tourism, and the dollars that a unit designation would bring. But critics argue that even the jewels of the national park system, such as Yosemite, Yellowstone, and the Grand Canyon, are falling into disrepair or suffering from a lack of staff in order to keep open areas that might better be served if privately protected and operated.

Protected area conflicts are usually not as well publicized as other types of natural resource controversies. While parks still have plenty of visitors, both from the United States and other countries, there are few nongovernmental organizations that can raise the amount of money the parks need. Even fewer are the number of lobbyists in Washington, D.C., who will fight for an entire system in need. Pork-barrel politics and additional designations are more likely than additional federal funding for the country's national parks. Without an identifiable political base or constituency, protected areas with fewer visitors that have less of an economic impact on an area, may find their resources plundered, sold, or lost altogether.

# Rangeland Management

Cattle ranching in the United States is a product of an influx of Spanish missionaries in what is now Texas, who let their long-horn cattle run freely across the open range. In 1830 an estimated 100,000 cattle roamed in Texas, and by 1860 the number had grown to 3.5 million. In the mid-nineteenth century, cowboys began to round up the wild cattle and drove them to areas where they could be sold. The industry began to flourish as new breeds were introduced, and grazing areas were free and seemingly endless. At the time, the vast publicly owned lands of the West were virtually independent of any type of government regulation. There was little understanding of, or interest in, the value of rangelands as a natural resource.

The General Land Office, established in the Department of the Treasury and transferred to the Department of the Interior in 1849, took on responsibility for western expansion and settlement, and grazing practices gradually began to change as more people came to the West. The 1862 Homestead Act encouraged settlers to identify a parcel of land, up to 160 acres, for which they received a patent. The patent owners could use the land for any purpose: mining, grazing, farming, or cutting timber. Homesteaders interpreted the law as meaning that they could settle on their 160-acre parcel and then graze livestock on adjacent, unclaimed lands. The custom of the time was based on the phrase "He who owns the water owns the land"; fencing was nonexistent. In 1878, explorer John Wesley Powell, who suggested that unirrigated grazing parcels be available in units of 2,560 acres (four square miles), also recommended that Congress standardize grazing rights using procedures similar to those of the 1872 General Mining Law. Ranchers would claim public lands through a patent system, with appropriation tied to the ownership of watering places. The plan, Powell argued, would legitimize the rights of permanent settlers who had "legitimate" control of water frontage, and thus the right to adjacent grazing land, over transient stockmen who grazed the open range in the absence of definitive property laws. Claims were backed by fences erected by individual ranchers, creating preemptive range rights (Hage 1994).

Powell's proposals did not gain sufficient support in Congress, and by 1885 an antifencing law was enacted that allowed the federal government to remove any unlawful enclosure. Powerful Eastern interests, who saw the natural resources of the West in terms of dollar signs, quickly bought up parcels in 160-acre increments. State laws governed water, along with the doctrine of prior appropriation, which was interpreted as meaning that the first user of a stream in time became the first user in right as well. The U.S. Supreme Court backed up that view in the 1890 case *Buford v. Houtz*, ruling that preemption laws gave settlers the right to graze cattle on the prairies, and by custom, the right to water their livestock on the public land. Forest Service chief Gifford Pinchot was heavily criticized for proposing that livestock owners begin paying fees to graze their animals within the national forests, even though the fee would have been about five cents per animal unit month (AUM). An AUM is the amount of forage (usually about 250 acres of grassland) necessary to sustain a cow and a calf for one month. Ranchers rebelled against Pinchot's plan, arguing that the fees amounted to a tax. But in 1906, a grazing fee was instituted by the U.S. Forest Service; it applied only to forest reserve lands, not those in the public domain. The number of permits and acres devoted to grazing rose dramatically, even with the fees. Between 1908 and 1920, livestock grazing increased from 14 to 20 million AUMs.

Congress was caught between two conflicting issues. On the one hand was the heavy national debt incurred as a result of World War I, and the realization that grazing fees could be increased to better represent the fair market value of the land's use. Ranchers were expected to complain, but so many cattle and sheep were already being grazed on public rangeland that they would have little option but to pay the additional costs, which were still considerably lower than fees for grazing on privately held land. On the other hand were members of Congress, mostly from Western states, whose constituencies had traditionally expected the federal government to stay out of their way. Powerful ranching interests had long considered grazing on federal lands a right rather than a privilege, and they supported legislators who agreed with them.

The result was a classic battle between East and West. Rancher and senator Robert Stanfield (R-OR), who was also a grazing permittee, orchestrated efforts in Congress to hold hearings on the fee system and rangeland management. Livestock companies rallied behind him, fully expecting that their interests would prevail, as they always had. But professional organizations, such as the Society of American Foresters and the American Forestry Association, gained the support of the press. Harsh editorials were printed in Eastern newspapers about the greed of the ranching industry and their assumption that they were privileged when it came to the use of the public land. Ranchers struck back by resisting proposals to force them to pay grazing fees not only within the national forests but also on the rest of the public domain (Layser 2006).

## Regulating the Range

Today, grazing policy stems from the 1934 Taylor Grazing Act, enacted to provide economic stability for Western ranchers by creating organizational arrangements to manage livestock grazing more efficiently without harming ranching interests (Davis 2001). The statute allotted qualifying ranches, termed base properties, an exclusive number of annual AUMs to graze their livestock on federal public lands. Fees were initially kept low, and field rangers kept conflict at a minimum by handling disputes informally. The Grazing Division, created under the statute, was responsible for administering the law; the agency was renamed the Grazing Service in 1939, and it merged with the General Land Office in 1946 to create the Bureau of Land Management. Few changes in statutes or policy were made until 1952, with passage of the Independent Offices Appropriation Act. That statute provided that when federally owned resources or property was leased or sold, a fair market value should be obtained, a provision that has rarely been applicable to public/private transactions.

Today an estimated 257 million acres of the total 383 million acres of federal public land in the continental United States are managed for grazing by four agencies: the Bureau of Land Management (BLM: 163.3 million acres); the National Park Service (NPS: 3 million acres); the U.S. Fish and Wildlife Service (FWS: 1.4 million acres); and the U.S. Forest Service (FS: 89.5 million acres). About 18 million AUMs are leased from the federal government; nearly two-thirds of federal public lands in the Lower Forty-eight states is now being grazed. The majority of the leased land is in eleven Western states:

1. Arizona        8.4 million acres
2. California      75,000 acres
3. Colorado       2.6 million acres

| | | |
|---|---|---|
| 4. | Idaho | 1.9 million acres |
| 5. | Montana | 4.1 million acres |
| 6. | Nevada | 110,000 acres |
| 7. | New Mexico | 8 million acres |
| 8. | Oregon | 550,000 acres |
| 9. | Utah | 3.15 million acres |
| 10. | Washington | 873,000 acres |
| 11. | Wyoming | 3.6 million acres (National Public Lands Grazing Campaign 2005) |

Over the years, grazing conflicts have arisen on several fronts. On one side of the current disputes are ranchers, livestock owners, and the organizations that represent their interests. One of the largest groups is the National Cattlemen's Beef Association (NCBA), founded in 1898. This trade association is a federation of state beef councils that coordinate efforts to build consumer demand for beef, as well as advocating for the cattle industry's political interests. The public's demand for beef had declined prior to the 1980s, when the industry initiated an extensive $50 million marketing campaign. Beef was losing out to Americans' preference for chicken, pork, seafood, cheese, and vegetables. In addition, imported beef, primarily from Japan, Brazil, Mexico, and Canada, has cut into the domestic share of beef being sold, further decreasing profits. Outbreaks of disease, from mad cow scares to instances in which restaurant hamburger was infected with *E. coli* bacteria, affected consumer buying habits.

Another perspective is provided by landowners who argue that a grazing lease actually is the legal recognition of a preexisting right to graze on the land. The Taylor Grazing Act made clear that the public lands were to be leased, not sold, and that the grazing permits did not create "any right, title, interest, or estate in or to the lands." That language was interpreted by the U.S. Supreme Court in *U.S. v. Fuller* in 1973, in which the Court ruled that the provisions of the Taylor Act made clear the congressional intent that no compensable property right be created in the permit lands themselves as a result of the issuance of the grazing permit. Property rights groups have been unsuccessful, for the most part, in gaining support for this position, although one rancher, Wayne Hage, has pushed his case through the federal courts for more than a decade.

On the other side are environmental groups whose opposition to the livestock industry focuses on the ecological impacts of grazing, including concerns about water quality, streambank stability, soil degradation, the increasing number and spread of invasive plant species, and the loss of natural forage for other wildlife. They point to several problems created by public lands grazing, although there is no consensus on how best to solve the conflict.

## Subsidization of the Livestock Industry

One of the major contentious issues is the subsidization of the livestock industry by the federal government directly, and by taxpayers indirectly. Currently, ranchers pay $1.43 per AUM to graze their animals on publicly owned lands, while the cost of grazing livestock on privately held land in the West ranges from $11 to $12 per AUM, according to the American Lands Alliance (ALA). Critics call the subsidies "welfare ranching," because grazing on public lands costs considerably less than current market prices. "For many years, the federal grazing fee has been set at [a price that is] less than it costs to feed a gerbil for a month" (Rosenberger 2004).

The current grazing lease programs cost an estimated $500 million per year, with only about $7 million per year collected from grazing fees. The ALA contends that more than half of federal grazing permits are leased to "hobbyists" who are not dependent upon ranching income, dispelling the myth of livestock roundup cowboys who have lived off the land for generations. They dismiss arguments that reducing or ending public land grazing will wipe out an entire culture in the region.

A study conducted by the GAO in 2004 showed that federal agencies lose at least $123 million a year keeping public lands open. Environmentalists contend that the figure does not include indirect expenses that are usually borne by local residents and taxpayers, including the introduction of noxious weeds and invasive species, sedimentation, and the negative effects on regional hydrology. A representative of the Western Watersheds Project says that the figure should be closer to $500 million per year in losses (Talhelm 2005).

Livestock trade associations argue that if the government does not help to pay the cost of grazing, beef prices will escalate rapidly, and the cost will be passed on to consumers. Another contention relates to the future of the livestock industry and the impact on local communities. One 1986 survey of ranchers reported that, when asked what they would do if prohibitions were put on public grazing, 21 percent reported that they would retire, 16 percent said that they would find a new occupation, 9 percent said that they would move to another state, and 21 percent said that they would sell their land to private developers (Anon. 1996). The NCBA also contends that ranchers are good stewards of the land because they depend on natural resources for their livelihood. They liken the health of the land to good business practices; when the land is more productive, ranchers thrive.

## Environmental Impacts of Grazing

The approach taken by many environmental organizations has been to emphasize the ecological damage done by livestock, especially on sensitive areas. Dramatic photographs and video footage of land before and after cattle have grazed an area are a potent argument for public consumption. Pictures of grassland on one side of a fence where grazing has been prohibited are in stark contrast to the dusty and compacted soil where hooves and manure have degraded the land. The use of such imagery on websites, at public forums, and at legislative hearings is a key strategy of many of the most prominent antigrazing groups.

The High Country Citizens' Alliance (HCCA) is attempting to differentiate between historical overuse of public lands and current overuse. In the past, they note, the public lands were treated as private lands, and some livestock owners either did not know about, or did not care about, the consequences of overgrazing. Even with a reduction in the number of livestock being raised today, past damage has created ecological conditions that may take hundreds or thousands of years to return to a healthy state. That is why, they say, it is important not to repeat past mistakes, or to compound them by continuing practices now known to be damaging to ecosystems in the West. "No single grazing system is a panacea in every location," they note. "We embrace neither cows nor condos, but seek middle ground that speaks for the land. The language of that middle ground is grazing reform" (High Country Citizens' Alliance 2005).

Other organizations, such as Public Lands Ranching, point to the increasing conflict between ranchers and recreationists, who often ignore fencing, gates, and signage and enter livestock grazing areas with little regard to the animals. It is not uncommon for a rancher to discover that water tanks and pipelines have been shot at, gates have been left open, and fences need costly repair because of carelessness or vandalism.

RangeNet, for instance, is a network of individuals created in 2001 as a semiautonomous project of the group Western Watersheds Project, Inc. Based in Beaverton, Oregon, RangeNet notes that while no other use impacts as many acres of Western public lands as frequently as grazing by domestic livestock, those impacts often cannot be detected by the untrained eye. Although some damage is easy to see, such as forage that has been grazed down to almost nothing, other effects are more subtle, including harm to hydrological cycles, indigenous organisms, and water quality. The group focuses on facilitating communication among rangeland activists (whose membership is by nomination and invitation only) and often uses photographs on its website.

Groups like Forest Guardians, based in Santa Fe, New Mexico, have approached the rangeland management issue by monitoring closely the actions of federal agencies such as the BLM and FS to show the ecological impact of livestock grazing, especially in riparian (streamside) areas that are prime habitat for wildlife. They allege that the two agencies failed to comply with routine environmental standards to limit environmental damage caused by cattle grazing on 50 to 75 percent of all grazing allotments in Arizona and New Mexico between 1999 and 2003 (Forest Guardians 2005).

## Conflict Management on the Range

Most of the efforts to reduce rangeland conflict have been small-scale pilot projects rather than state- or national-level initiatives. In southwest Idaho, for instance, a U.S. district judge in Boise ordered the owners of fourteen ranches to remove all of their cattle and sheep from about 800,000 acres of federal land that is part of the Jarbridge Resource Area. The land is managed by the Bureau of Land Management, and several lawsuits have been filed against the agency for failing to complete adequate assessments of the damage done by livestock. Opponents of grazing argue that overgrazing has harmed both the landscape's vegetation and wildlife habitat, as well as degrading stream water quality.

In an effort to stem the tide of lawsuits and continued grazing, one activist, John Marvel, proposed that the owners of four ranches reduce the number of cattle they graze on the disputed land while a comprehensive environmental impact study is completed. If the study finds that the overgrazing is still taking place, then the BLM could impose even more restrictions on grazing permits. In the meantime, not all cattle would need to be removed, the court battles could be placed on hold, and the habitat would be allowed to recover, Marvel says (Ring 2005, 6). Another project involves ranchers in Hot Springs Canyon, along the San Pedro River in southeastern Arizona. In 1988 a group of friends, many of whom had backgrounds as human rights activists, purchased 135 acres of land and formed the Saguaro Juniper Corporation. They hoped to bridge the divide between ranchers and environmentalists by forming a consensus-based community in which they could successfully raise cattle while embracing management techniques that focus on riparian and rangeland restoration and improvement. At the heart of the group is a belief that an organic process of societal interaction is possible in ranching, rather than a traditional corporate farm orientation. Among the strategies used are rotating grazing areas, resting the land during the summer when grasslands are in their peak growing season, and the humane treatment of animals. By 2005 the group had grown to more than sixty shareholding "associates" who own over 1,000 acres of land, 8,000 acres of state grazing leased land, and 750 acres of privately leased land. In addition to practicing new types of grazing management, the corporation's associates participate actively in the community and cooperate with governmental agencies and other organizations.

Montana ranchers have formed an alliance, the Blackfoot Challenge—"because that's what it is, a challenge"—to bring together governmental agencies, timber companies, and conservation groups in a community effort to protect natural resources in the Blackfoot River Valley. By using tools such as conservation easements and buyout incentives, and with a budget of $2.8 million, the grassroots initiative has already protected 90,000 acres of private land from development and restored 47 miles of stream, 2,600 acres of wetlands, and 2,300 acres of native grasslands. The federal government has assisted by sharing the costs associated with grazing management and developing off-site water sources for cattle, allowing for the restoration of riparian areas. Conservation groups consider the effort a way of creating corridors for wildlife while bringing together 500 private landowners, four private corporations, thirty-six governmental organizations, nine foundations, and fourteen conservation and environmental organizations. The initiative helped ranchers to avoid potential conflicts on their own, without being forced to do so by the federal government (Little 2005).

At the national level are organizations that promote a more controversial strategy: the voluntary buyout of grazing leases. What differentiates the groups are the logistics of trying to convince ranchers to end what some consider to be abusive grazing practices by amending existing federal laws to permit those with grazing permits to sell them, either to the federal or state government or to an organization that would retire the lease. Cooperative agreements and partnerships have been moderately successful, but there are examples in which they have been met with substantial resistance by local and state officials.

In 1998, the Arizona-based Grand Canyon Trust began paying ranchers in Utah to give up their grazing rights near the Escalante River. The land is marginal economically and at risk environmentally, according to the trust. Ranchers could consolidate their herds in areas where there was more forage for their livestock, and be paid by the trust so that the land could be rested and restored. The program was especially beneficial to ranchers who were losing money on their livestock operations. The organization, which is dedicated to the preservation of the Colorado Plateau, had paid more than $1 million and ended grazing on more than 400,000 acres by 2005.

Although the plan seemed to benefit both ranchers and conservationists, one of Utah's state representatives, a former BLM employee, organized local officials to roll back the agreements that had been forged by the Grand Canyon Trust. The Utah residents argued that by reducing the number of acres being grazed, the organization was hurting the ranching culture of the region and depriving young people of the "character-building chance to work on the land." Trust officials were angered that their free-market approach to restoring the land was being thwarted. "We've been out there dealing with this. We solved the problems of the BLM and we're hurting the Kane County economy by buying out guys who are going bankrupt? I don't get it" (Barringer 2005).

The National Public Lands Grazing Campaign (NPLGC), composed of a coalition of Western antigrazing groups, has perhaps the most ambitious proposal under consideration. The group has a threefold mission: to educate the American public about the destructive impact of livestock grazing from an environmental and economic perspective; to enforce existing grazing laws and regulations; and to allow voluntary, permanent buyouts of grazing leases on public lands at the rate of $175 per AUM. The organization estimates that livestock grazing can be ended altogether for an average of $13.45 for each acre in the program.

One of the factors that has made it difficult to come to any agreements on rangeland management is the division among environmental organizations as to which strategy to follow. The NPLGC, for instance, is focused almost exclusively on its permit buyout proposal. The Sierra Club, also actively involved in the policy debate, advocates significant changes to current land management practices but also acknowledges that grazing may achieve some ecological objectives. The group also notes that there are areas where settlement or ownership patterns may require grazing managers to move toward the restoration of the land, native plant and animal communities, and water quality on an incremental basis. The NPLGC counters that many public lands livestock operators could not meet the Sierra Club's management goals because the majority already are facing financial difficulties. Trying to get these ranchers to engage in less environmentally abusive practices "will render their grazing operations untenable from what was previously merely unprofitable" (National Public Lands Grazing Campaign 2005).

Rangeland conflict is not likely to be addressed or solved by governmental pronouncements, Washington lobbyists, or protest groups. The custom and culture of the West is so thoroughly ingrained in U.S. history that it will be difficult for grazing practices and policies to change without consensus and collaboration among the stakeholders. But that is also an issue where nonlegalistic solutions are most likely to be successful, as more and more ranchers recognize that conditions on the range, and the economics of livestock ownership, have already changed dramatically.

# Timber and Forests

The overview of the history of forest management and conflict in Chapter 1 provides a background for this section, which goes into more depth in exploring specific issues. Not surprisingly, many of the problems that existed in the nineteenth and twentieth centuries are still around in the twenty-first century. This does not mean that conflicts cannot be resolved, and, in fact, there have been many successful partnerships, collaborative agreements, and pilot projects in timber disputes. But this is a policy area in natural resources in which the successes have tended to be on a smaller, more localized level. Deep-seated, emotional perspectives still divide many of the key stakeholders in these conflicts.

What follows is an explanation of two of the most contentious problems in the United States regarding forests as a natural resource: maintaining a sustainable supply of timber, and protecting old growth forests. These issues are somewhat overlapping, since protection of forests was a part of the nation's earliest environmental policies. But the way in which conflicts related to them have been resolved does differ substantially from other natural resource issues analyzed previously in this chapter.

## Maintaining a Sustainable Supply

With the creation of the U.S. Forest Service in 1905, Congress gave its blessing to the founding principle that had been the vision of its first chief, Gifford Pinchot: "The continued prosperity of the agricultural, lumbering, mining, and livestock interests is directly dependent upon a permanent and accessible supply of water, wood, and forage, as well as upon the present and future use of these resources under businesslike regulations, enforced with promptness, effectiveness, and common sense" (Wilkinson 1992, 128).

Pinchot's European training had shown him the need for changing forest practices, and he fully intended to make sustainability the dominating forest management goal for the twentieth century. In forest parlance, sustainability means that, under federal law, management agencies must harvest trees at a specific pace. By the time the last tree of the original, virgin forest is cut, the first tree of the regrown forest should be big enough to harvest. Maintaining sustainability is essential for the "permanent and accessible supply" that Pinchot sought as a replacement for the slash and burn practices of the past. These methods, pioneered in Germany and France, would bring the young country up to date with ideas being used elsewhere.

One of the practices that Pinchot considered among the most wasteful of trees as a resource was clear-cutting—logging all of the trees within a given area without regard to the value of the lumber produced. His philosophy of utilitarianism was based on the principle that forests provide more products than can be used efficiently. They are wasted through death and decay but should not be exploited by humans. Clear-cutting regained popularity in the late 1940s and early 1950s, even though so much of the timber ended up being left to rot on the ground or being burned in huge slash piles. Clear-cutting was less expensive and more efficient than selective logging, since it did not take the extra time required to select and then thin out the desired number of trees. Heavy mechanized equipment made quick work of acre after acre as logging trucks formed their own highways through the national forests.

During the late 1960s, a combination of high demand and rising prices made clear-cutting even more desirable. Getting out the cut by logging as extensively as possible became the norm in the national forests until the early 1970s. A 1973 lawsuit by the West Virginia division of the Izaak Walton League against the secretary of agriculture, Earl Butz, alleged that the use of clear-cutting in the state's Monongahela National Forest was in violation of the Organic Act of 1897. The Forest Service countered with their interpretation that the statute directed the agency to manage the national forests scientifically, leaving the choice of practices to the discretion of the department. The major goal of the Forest Service was to maintain a continuous supply of timber.

The Forest Service lost the initial court battle in U.S. District Court, and the opinion made it clear that the agency was allowed to sell only timber that was dead, mature, or large, and it was required to mark each individual tree for sale, rather than clear-cutting. The government appealed the case, and meanwhile, an almost identical case was filed in the Tongass National Forest. In the latter case, the plaintiff did not rely upon the Organic Act but, instead, cited NEPA and a series of other environmental statutes. The Monongahela case was affirmed by the U.S. Court of Appeals, while the U.S. District Court in Alaska ruled that the Forest Service did not violate the statutes cited by the plaintiff but was in violation of the Organic Act, citing the West Virginia ruling as precedent.

The two cases were important not only because they formed the legal precedent against clear-cutting: they also demonstrated the power of citizen groups to use litigation as a strategy. Although litigation had worked, it was time consuming and costly. The Forest Service countered by seeking timber industry support in Congress, where there appeared to be more support for increased logging. The National Forest Products Association proposed legislation that would substantially increase timber yields, and after some initial confusion over what action to take, environmental organizations, led by the National Wildlife Federation, the Sierra Club, and the Wilderness Society, mounted a vocal opposition to the proposed statute. The Forest and Rangeland Renewable Resources Planning Act of 1974 (RPA) called for a nationwide system of forest planning. It was followed in 1976 by the National Forest Management Act (NFMA), which became an amendment to the RPA. The NFMA repealed the controversial section of the Organic Act by eliminating the restrictive language referring to dead, mature, or large trees. Most important, NFMA provided a mechanism for public participation and scrutiny, including public hearings, prior to timber sales. It forced the Forest Service to develop management plans for each of the nation's 155 national forests. Although the statute did not limit clear-cutting, it did resolve at least some of the conflicts over forest management by requiring the inclusion of scientific studies in preparing forest plans.

By January 1990, the Forest Service had developed plans for all but one of the national forests, and ninety-eight unit plans had been finalized. Almost all were immediately appealed by environmental organizations, and several were litigated, as opponents again turned to the courts for a remedy. It became clear that the sustainability issue could not be resolved easily or quickly, as groups like the Wilderness Society and Sierra Club held their ground. Over the next fifteen years, appeals and litigation, under both NEPA and the NFMA, became commonplace. The likelihood that an appeal would be filed caused the Forest Service, in some cases, to cancel plans for timber sales because of the resources that would be required to justify any logging at all. Several Native American tribes were so opposed to logging on tribal land that the Forest Service stopped all proposed sales, instead creating a new position within the agency to deal specifically with individual tribes (Vaughn and Cortner 2006).

Another major statute that affected timber supplies was the Multiple Use and Sustained Yield Act of 1960 (MUSYA). Congress identified five purposes that the national forests were designed to serve: outdoor recreation, range, timber, watershed, and wildlife and fish. Multiple use was well understood; what was more controversial was the concept of sustainable yield in the era of environmental activism. The law stated that sustainable yield referred to the achievement and maintenance in perpetuity of a high-level annual or regular periodic output of the various resources of the national forests without the impairment of the productivity of the land.

Conflicts over sustainability and the need to provide for the growing demand for timber did not end with the passage of the NFMA, however. Just a few years later, another resource issue would emerge that moved government officials and environmental organizations in another direction altogether—protecting the habitat of a rare bird.

## Protecting Old Growth Forests

For much of the nation's history, the forest was considered an obstacle because it blocked forward movement and took up valuable agricultural land needed for ranching and farming. One account of a stage journey through the Pacific Northwest around 1865 described "trees larger and taller and standing thicker; so thick and tall that the ground they occupy could not hold them cut and corded as wood. Washington Territory must have more timber and ferns and blackberries and snakes to the square mile than any other state or territory of the Union" (Dietrich 1992, 20). Progress meant cutting down the forest, if only to get through the trees to what lay beyond. In 1897, President Grover Cleveland had set aside two-thirds of the Olympic Peninsula as forest reserves because of fears of a "timber famine." The timber industry protested, and the reserve was reduced by a third until President Theodore Roosevelt restored the reserve as a 620,000-acre national monument in 1909. In 1915 the monument's acreage was cut by President Woodrow Wilson, and its status changed again when President Franklin Roosevelt named the area a national park. One of the reasons for the back and forth debates over this particular tract of land was the size of its trees.

A 1902 government survey of Washington's Olympic Peninsula called the region the most heavily forested region of the country, but because of its remoteness, it was one of the last areas to be logged. The trees in the Pacific Northwest are conifers such as Douglas fir, western hemlock, and Sitka spruce, and eventually many of the largest, most valuable trees were cut and replaced by an orderly, industrial forest. For decades the lack of roads kept harvests to a minimum until World War I, when a railroad system was put in place to provide wood for the war effort. By that time few of the big trees were left, and they formed a mosaic between the mountains and the ocean.

In the Pacific Northwest, the timber industry was the life-blood of many small communities, for generations of families. In 1985 logging accounted for 4 percent of the workforce in western Oregon and 20 percent of the area's total manufacturing sector employment; in 1988 the Forest Service estimated that 44 percent of Oregon's economy and 28 percent of Washington's economy were directly or indirectly dependent on national forest timber. The stakeholders went beyond just timber companies, however. The area was home to many smaller, family-owned mills, and log haulers were well paid for their work in an area that had few highly paid jobs. Local governments also were deeply involved in timber harvests; 25 percent of the revenues from timber harvested on federal lands is returned to municipalities where the forest is located, to be used for schools and roads. Federal timber revenues accounted for between 25 and 66 percent of the total income for ten Oregon counties during the late 1980s (Layser 2006). Any proposed reductions in harvests meant deep cuts in city and county services in many rural areas.

In California, the situation has been slightly different. The species that has been the subject of controversy is the coast redwood, covering about 1.7 million acres along the state's northernmost coast. About 350,000 of those acres are in the public domain, with the remainder owned by several private timber companies. Redwoods are not considered threatened, because they have a shallow root structure that grows very quickly. The conflicts over the coast redwoods are over *old growth,* a term that usually refers to trees 250 years old or more, with a trunk diameter of four feet at breast height (dbh). There are an estimated 90,000 acres of old growth left in northern California, and about 80,000 acres are currently preserved by the federal or state government.

By the mid-nineteenth century, loggers had harvested most of the ancient forests in the state, some containing trees more than 2,000 years old. By the end of the twentieth century, almost all remaining old growth was found in the national forests, and, symbolically at least, that means that the trees belong to all Americans. But the goal of the FS and state resource agencies—maximizing income from logging—conflicted with the desire of environmental groups: to save those trees more than others. By the 1980s forests were being cut much more rapidly than they were being restored, and old growth trees provided substantially more lumber than younger, smaller-diameter ones. Environmental organizations were concerned about predictions that all of the trees over 200 years old would be gone in less than thirty years (ibid.). The battle over old growth in Washington, Oregon, and California became a very real conflict over aesthetics, economics, and the habitats of endangered species, pitting small communities and large logging interests against environmentalists who vowed to do whatever it would take to protect the trees.

The difficulty with environmental groups' strategy prior to the 1990s was that old growth forests were considered a regional, rather than a national, issue, making it more difficult to get support in Congress for protective legislation. According to forest activist Andy Kerr: "Expecting the Northwest congressional delegation to be rational about ending the cutting of ancient forests in the late 1980s is like expecting the delegation from the American south to deal rationally with ending segregation in the late 1950s" (Davis 2001, 68). As the issue gained public exposure and political salience, protecting old growth became a politically defensible position to take.

Transposed upon the old growth debate was a biodiversity issue that overshadowed the logging of timber. Under the 1973 Endangered Species Act (ESA), the Forest Service was required to protect the habitat of the Northern spotted owl, which nests and lives in ancient forests in the Pacific Northwest. For more than a decade, studies were made of the bird's nesting habits and range to determine whether it would be subject to the ESA. In the meantime the agency did as little as possible in an attempt to follow direction from the chief of the Forest Service that whatever arrangement was finally agreed upon, the decision would not affect allowable timber sales in the region by more than 5 percent (Dombeck, Wood, and Williams 2003).

Both Oregon and Washington gave the spotted owl protected status under state law in 1988, putting additional pressure on the federal government to take similar action. But efforts to develop a conservation strategy that would please all of the affected stakeholders became heavily politicized, even after lengthy congressional hearings and a scientific report recommending a managed landscape to protect entire habitat areas rather than nesting sites. In June 1990, the U.S. Fish and Wildlife Service declared the bird "threatened," a less protected status than "endangered," but one that resulted in large sections of national forest lands being declared off limits to logging. Subsequently, environmental organizations sued the government for delaying the ESA process and for failing to comply with NEPA and the NFMA. A federal judge agreed, and timber sales were temporarily halted.

In April 1993, President Bill Clinton attempted to resolve the spotted owl controversy by inviting timber workers, community activists, and representatives of environmental groups to Portland, Oregon, for a national Forest Summit. No officials from the Forest Service or the Bureau of Land Management were invited to speak. The result was a presidential decree for the agencies to come up with legally defensible plans for the owl and for the future management of old growth forests in the Pacific Northwest. In 1995, an amendment called a rider was attached to the Forest Service appropriations bill for fiscal year 1996 that allowed all salvage timber sales to be temporarily exempt from legal challenge or judicial review. What the Clinton administration considered a reasonable compromise only mobilized environmental groups against the plan. Once again they turned to the courts, filing appeals against virtually every project the Forest Service proposed. Sales since then have plunged as cases make their way through the judicial arena, with the logging industry accusing the environmental movement of extremism, massive job losses, and the end of the timber industry in the Pacific Northwest.

In California, however, the old growth issue has been more emotional and symbolic than legalistic and economic. The flashpoint came in 1986, when MAXXAM Corporation acquired Pacific Lumber Company (PLC). Prior to the acquisition, even critics of PLC felt that the company was making efforts to harvest trees sustainably on the land it owned, without resorting to clearcutting. MAXXAM's owners had different plans, and sought to increase substantially the number of acres logged each year, using intensive harvest methods in the old growth Headwaters Forest. Those plans enraged environmentalists, from the radical, direct-action group Earth First! to regional organizations such as the Environmental Protection Information Center. More than 1,000 people were arrested at a protest in 1996, and 9,000 activists rallied against the company in another protest in 1997, the same year that Julia "Butterfly" Hill began a two-year-long tree-sit in a tree she named Luna. In 1998 a small group of activists chained themselves to shelves at a Home Depot home improvement store near Santa Rosa to protest the sale of rakes and brooms whose handles were said to be made out of old growth. Organizations such as the Campaign for Old Growth, Forests Forever, and the Save the

Redwoods League rallied around the idea that all remaining coast redwoods represent a national treasure that should be placed under a public trust for protection.

The federal government sought to resolve the Headwaters conflict by negotiating an agreement among the Bureau of Land Management, the state of California, the PLC, and the Elk River Timber Company—which owned lands adjacent to those of the PLC—to purchase the most sensitive areas for $380 million. Under the terms of the agreement, finalized in March 1999, the PLC would be permitted to log on the remaining 200,000 acres it owned. The agreement required the company to develop a complex habitat conservation plan for another endangered bird, the marbled murrelet. The new 7,400-acre Headwaters Forest Reserve solved some problems by providing protection for some old growth trees, but not all of them. Many political leaders and some environmental groups considered the agreement to be better than going through years of legal wrangling, as had been the case in Oregon and Washington, while logging continued. Other activists felt that the buyout had simply given the PLC the go-ahead to continue logging old growth on other land the company owned, and there was skepticism about whether the murrelet and other sensitive species would really be protected.

Simply put, the Headwaters purchase agreement did not end the conflict. In 2000 the California Board of Forestry considered a temporary ban on cutting any old growth trees, and in 2001 a coalition of conservation groups submitted a state ballot initiative that would have prohibited private land owners from logging or damaging old growth trees that were alive in 1850, the year in which California became a state. One of the initiative's sponsors claimed that "we are not only saving trees. We are saving a bit of our humanity by taking responsibility as stewards of God's creation. This is not merely an ecological issue, but a moral and ethical issue" (Campaign for Old Growth 2006). The initiative did not gain enough signatures to be placed on the November 2002 ballot. In 2004 the California legislature considered the Heritage Tree Preservation Act to protect old growth, but it, too, failed to get sufficient support for passage.

## Collaboration Efforts

While old growth controversies have been marked by lawsuits, protests, and political brokering, there have been several successful collaborations among stakeholders to try to deal with the fact that "the 'timber wars' damaged everybody and served nobody's true interest" (Quincy Library Group 2005). In 1992 a timber industry forester, an environmental attorney, and a member of the county board of supervisors began informal discussions that would lead to the creation of the Quincy Library Group (QLG). The initially informal organization dealt with the issue of forest management within a three-county area in northeastern California with a population of about 50,000. The area is heavily dependent upon the logging industry, has extensive hydroelectric facilities, and is popular with recreational users—in short, all the elements needed for natural resource conflicts.

The group took its name from its meeting place, and a steering committee of about thirty people began to map out a proposal for integrating forest management and the community's economic needs with biodiversity protection and fire and fuels management objectives. The group developed a northern Sierra working circle, recommendations for timber management, expanded stewardship agreements, strategic fuel reduction to reduce wildfire risk, guidelines for riparian area protection, the removal of acreage from road construction and timber harvesting, and selective tree thinning. The U.S. Forest Service received funding to implement some of the QLG's proposals as part of a forest health pilot program, and, in conjunction with the National Renewable Energy Laboratory, the organization cosponsored a study of the feasibility of manufacturing ethanol from small-diameter trees (ibid.). Similar collaboratives have been established in Oregon and Washington, most notably the Applegate Partnership in southern Oregon.

While some organizations still seek a "no cut" policy on public lands, most of those involved in a collaborative approach realize that compromise is the only real answer. Groups may be able to win some battles over old growth, but it is doubtful that they will ever end logging altogether. The political clout of the timber industry, coupled with a steady demand for products such as lumber and paper, will help keep Pinchot's promise of a permanent and accessible supply of timber alive.

### References

American Hiking Society website at www.americanhiking.org (accessed 2006).

American Lands Alliance. "Overview"at www.americanlands.org (accessed November 6, 2005).

Anon. 1996. "Cattle Grazing on Public Lands: The Hard Fought Battle in the Southwestern United States." *Beckoning* (October 5) at www.thebeckoning.com (accessed August 17, 2005).

Barringer, Felicity. 2005. "Plan to Restore West Grasslands Meets with Resistance in Utah." *Arizona Republic*, December 4, A19.

Burkhart, Michelle. 2005. "Public Lands Agenda Turns More Radical, Urgent." *High Country News* 37, no. 22 (November 28): 5.

Cable, Sherry, and Charles Cable. 1995. *Environmental Problems, Grassroots Solutions: The Politics of Grassroots Environmental Conflict.* New York: St. Martin's.

Campaign for Old Growth website at www.ancienttrees.org (accessed 2006).

Davis, Charles. 2001. *Western Public Lands and Environmental Politics.* 2d ed. Boulder, CO: Westview.

Dietrich, William. 1992. *The Final Forest: The Battle for the Last Great Trees of the Pacific Northwest.* New York: Simon and Schuster.

Dilsaver, Lary M. 1994. *America's National Park System: The Critical Documents.* Lanham, MD: Rowman and Littlefield.

Dombeck, Michael P., Christopher A. Wood, and Jack E. Williams. 2003. *From Conquest to Conservation: Our Public Lands Legacy.* Washington, DC: Island.

Forest Guardians. 2005. "Activists Take on Irresponsible Government." Website at www.fguardians.org (accessed November 6, 2005).

Freemuth, John. 1991. *Islands under Siege: National Parks and the Politics of External Threats.* Lawrence: University Press of Kansas.

Hage, Wayne. 1994. *Storm over Rangelands: Private Rights in Federal Lands.* Bellevue, WA: Free Enterprise.

High Country Citizens' Alliance website at www.thinairweb.com (accessed August 17, 2005).

Jenkins, Matt. 2005. "The Final Energy Frontier." *High Country News* 37, no. 23 (December 12): 7.

Klyza, Christopher M. 1996. *Who Controls Public Lands? Mining, Forestry, and Grazing Policies 1870–1990.* Chapel Hill: University of North Carolina Press.

Lay, Jennie. 2005. "Congress Bets on Oil Shale." *High Country News* 37, no. 23 (December 12): 8–9.

Layser, Judith. 2006. *The Environmental Case: Translating Values into Policy.* 2d. ed. Washington, DC: Congressional Quarterly Press.

Little, Jane Braxton. 2005. "Rising to the Challenge: Montana Ranchers

Practice Homegrown Conservation." *Nature Conservancy* 55, no. 4 (winter): 28–39.

Lowry, William. 2001. "The Impact of Reinventing Government on State and Federal Parks." *Journal of Policy History* 13, no. 4: 405–428.

National Park Foundation website at www.nationalparks.org (accessed October 18, 2005).

National Parks Conservation Association. 2005a. "Growth of the System: F" at www.npca.org (accessed November 1, 2005).

National Parks Conservation Association. 2005b. "National Park System Units" at www.npca.org (accessed November 1, 2005).

National Public Lands Grazing Campaign website at www.publiclandsranching.org (accessed November 6, 2005).

Public Employees for Environmental Responsibility. 2005. "National Parks to Seek Corporate Sponsorships." News release (November 30) at www.peer.org/news.

Quincy Library Group website at www.qlg.org (accessed 2005).

Ring, Ray. 2005. "Strange Bedfellows Make a Grazing Deal in Idaho." *High Country News* 37, no. 18 (October 3): 6.

Rosenberger, Jack. 2004. "Wasting the West: How Welfare Ranchers and Livestock Are Damaging Public Lands." *E Magazine* 15, no. 4 (July–August): 20–22.

Sellars, Richard W. 1997. *Preserving Nature in the National Parks: A History.* New Haven: Yale University Press.

Sierra Club website at www.sierra.org (accessed October 18, 2006).

Smith, Duane. 1987. *Mining America: The Industry and the Environment, 1800–1980.* Lawrence: University Press of Kansas.

Talhelm, Jennifer. 2005. "Report: Public Lands Grazing Costs $123 Million a Year." *Environmental News Network* website at www.enn.com (accessed November 1, 2005).

U.S. Department of the Interior. 2002. *Recreational Fee Demonstration Program.* Report No. 2002-I-0045. Washington, DC: U.S. Department of the Interior.

U.S. Department of the Interior, National Park Service. 2003. *National Park Service Fees: An Examination of Public Attitudes, Technical Report.* Flagstaff: Northern Arizona University, Social Research Laboratory.

U.S. General Accounting Office. 1998. *National Park Service: Efforts to Identify and Manage the Maintenance Backlog.* Report No. GAO-98-143. Washington, DC: General Accounting Office.

———. 2003. *Recreation Fees: Information on Forest Service Management of*

*Revenue from the Fee Demonstration Program.* Report No. GAO-03-470. Washington, DC: General Accounting Office.

Vaughn, Jacqueline, and Hanna J. Cortner. 2006. *George W. Bush's Healthy Forests: Reframing the Environmental Debate.* Boulder: University Press of Colorado.

Western Slope No-Fee Coalition website at www.sespewild.org (accessed October 18, 2006).

Wilkinson, Charles F. 1992. *Crossing the Next Meridian: Land, Water, and the Future of the American West.* Washington, DC: Island.

Williams, Ted. 2004. "The Mad Gas Rush." *Audubon* (March): 38.

# 3

# Worldwide Perspective

There is a substantive body of research related to sustainability when it comes to natural resources on a global scale. Political scientists, economists, sociologists, and ecologists can point to their studies of international agreements and treaties, scientific assessments, and efficiency. There is little disagreement about the basic goal: maintaining human security by respecting natural limits and promoting economies by managing overconsumption. The formula seems simple: improve, rather than degrade, the environment, and all people will prosper. Unfortunately, many of the scientific approaches to pollution have resulted in only marginal improvements. But increasingly, a second element has gained support: transformational change that focuses on reducing consumption of natural resources, rather than increasing production. This chapter looks at four issues that incorporate those principles, examining whether current problems and crises can be dealt with on a long-term basis, rather than by quick-fix solutions that rarely result in substantive change. Each section begins with an explanation of the scope of the problem, followed by an analysis of the nature of the conflicts that have taken place, and lastly, an overview of some of the attempts that have been made to resolve those conflicts.

While optimists might believe that the same types of conflict resolution strategies that work in the United States can be applied everywhere, (the "Why can't we all just get along?" approach), in reality, cultural, economic, and political differences among industrialized and developing countries make solutions much more difficult to find. The types of corruption and civil/ethnic warfare seen in some parts of the world limit how much cooperation and consultation can be attempted. Institutional barriers sometimes reduce the opportunities for public participation, as in the case of governments not built on models of representative democracy. Threats and intimidation have reduced the ability of some local activists to work toward compromise when their lives are at stake. But as is the case with any problem, building public awareness is the first step toward finding solutions.

# Diamonds and Precious Metals

In the United States, the history of mining is often romanticized, from the Gold Rush in California in the mid-1800s to the hearty prospectors looking for diamonds in the fields of Arkansas. The image of the old prospector and his mule is ubiquitous, even at a point in the twenty-first century when mechanized extraction is highly sophisticated. Whether the resource is coal, oil, or potash, the industry is regulated, and disputes are relatively small. Seldom do these types of conflicts over natural resources go further than the courtroom, and they definitely are no longer the source of violence. When political leaders argue, it is more likely to be over subsidies than ownership.

But in 2000, the World Bank called the struggle over diamonds and other mineral commodities the biggest cause of civil war globally, overtaking even political disputes. In Africa, rebel forces fight over mining areas and the trade in "hot rocks," or "conflict diamonds," is plundering natural resources to finance weapons and criminal activity (Masland 2000). Some analysts have gone so far as to question whether mining companies "produce" or even encourage wars to gain easier access to a country's mineral resources, although conspiracy theories about these types of conflicts are difficult to prove (Kennes 2005).

Somewhere between 4 and 15 percent of the world's $7 billion annual diamond business is illegal, and trade in these valuable stones, along with other precious metals, is fueling civil war in several African countries, including Angola, the Democratic Republic of Congo, Sierra Leone, Liberia, and Burkina Faso. Investigators have found evidence that rebel groups in some nations have used illicit diamonds, primarily those called "rough" because they have not yet been cut and polished into stones, in trade for weapons. Leaders in Sierra Leone's Revolutionary

United Front (RUF) are alleged to have traded more than 2,000 diamonds over a ten-month period, and the gems found their way into the marketplace. Local residents were often herded out of their villages in diamond-rich areas, or used as slave labor so rebel forces could mine the diamonds without interference. UN peacekeeper forces have been attacked and driven out of rebel-held areas, and Liberia has become one of the continent's centers for diamonds, drugs, arms, and money laundering, according to one report (Masland 2000).

In Angola, groups such as the National Union for the Total Independence of Angola (UNITA) rebel movement have battled with the Popular Movement for the Liberation of Angola (MPLA) since the 1950s and 1960s, when the countries were considered essential to maintaining the balance of power between the United States and the Soviet Union. The Cold War caused the two countries to support different rebel factions in an attempt to gain allies, but it had the effect of balkanizing nations, forcing them to compete more aggressively in the international trade market. Sanctions imposed against rebels have been ignored, and social disorder and the disruption of political authority have become the norm, leading to cross-border armed conflicts.

The "blood diamond" conflict has been accompanied by humanitarian tragedies and human rights abuses. Over time, the rest of the world is beginning to pay more attention to post-apartheid Africa, with the creation of international war tribunals, with intervention by the Economic Community of West African States and the African Union, and with assistance from the UN High Commissioner for Refugees. Although the civil wars are sometimes characterized as tribal or civil struggles, there is no longer any doubt that the conflict is financed by the sale of a major natural resource—diamonds (Orogun 2004).

There are many reasons why diamonds are the resource of choice in these conflicts:

- Diamonds are small, portable, and easy to conceal, making transport by courier or other methods relatively simple.
- Even rough (that is, unpolished) diamonds are extremely valuable.
- Diamond trafficking can be conducted without benefit of financial institutions, making it difficult to track the coming and going of funds.

- Bartering is a common and socially acceptable practice among rebel groups, who do not need to rely upon an industry intermediary to sell stones.
- Falsification of inventory records and transactions is relatively easy.
- Diamonds do not depreciate in value and are relatively immune from market fluctuations.

One of the reasons why diamonds have been at the center of so many conflicts is that most attempts to track them from mines to market have been unsuccessful. The diamond industry has tried to implement various strategies to make sure that blood diamonds are not traded openly, partially through labeling systems. One method is by determining the source of rough diamonds, commonly divided into the categories of "country of origin," meaning the place where the diamonds are taken from the ground, and "country of provenance," meaning the original location from which a particular parcel of diamonds in the transaction chain of reference came from. Experts believe that the immediate and preceding country may not actually be the country of origin, and that the critical link is finding out how the "provenance" came about.

Some companies, such as De Beers, have tried to classify diamonds in their possession (through the firm's Central Selling Organization) as "nonconflict" diamonds, so that their sales are not contaminated. But the company has also placed a date on acquisitions as being before or after March 26, 2004, with minimal accounting of where the rest of its current inventory originated. The company has also said that it would require guarantees from buyers that they were not dealing in conflict diamonds. However, it is increasingly difficult for a company to verify such claims or to authenticate where the diamonds may have traveled after they were mined. "Diamond laundering" and the process of relabeling gemstones to avoid detection and confiscation are not very difficult, since rough diamonds mined in different countries can be combined in mixed parcels with legitimately purchased rough stones (ibid.).

In the late 1990s, diamond-producing and -importing countries began their efforts to improve tracking of both rough and polished stones through a voluntary system of identification and inventory control. The industry began to realize that unless it was willing to police its own members, international regulation was likely to be more complicated and costly than they might want. The United Nations, recognizing that diamonds were being used for arms sales, successfully gained support for international sanctions against Angola in 1998, prohibiting imports of any diamonds that were not controlled through a Certificate of Origin program. Additional sanctions were later placed against Sierra Leone.

The current method of tracking diamonds, called a scheme, was launched in May 2000 after a meeting in Kimberley, South Africa. The United Nations held multiple meetings and conferences on conflict diamonds, and in December 2000 and again in March 2002, after approval of the Interlacken Declaration, the UN General Assembly passed resolutions calling for the development of an international system to control all diamonds, to keep conflict diamonds out of the legitimate gemstone marketplace. Members of the G-8 group of nations also called for an international system in 2000, identifying the trade in conflict diamonds as a key concern in conflict prevention. In June 2002, leaders also pledged support for addressing the link between natural resource exploitation and conflict in Africa (Natural Resources Canada 2006).

In the meantime the diamond industry took action as well, in part because of pressure from both governments and nongovernmental organizations (NGOs). On July 19, 2000, the World Federation of Diamond Bourses and the International Diamond Manufacturers Association issued a joint resolution in Antwerp, Belgium, calling for control of diamonds when they are still bought and sold as rough diamond parcels. The industry proposals called for all parcels to be sealed and registered in a universally standardized manner by an accredited export authority from the exporting country. Individual countries would be held responsible for creating diamond boards with the authority to seal parcels and to register them in an international database. Consuming countries would be prohibited from importing polished diamonds from any country that did not have a program in place to deal with rough parcels. The "diamond net" would also require countries, whether importers or exporters, to enact legislation to bring criminal charges against any individual or company proven to be knowingly involved with illegal rough diamonds.

South Africa now chairs a consortium of forty-four countries in the Kimberley Process International Certification Scheme. The members are responsible for 98 percent of the rough diamonds that are mined, produced, processed, imported, and exported in global trade. After intense and sometimes bitter negotiations, the Kimberley Process was implemented simultaneously among the participants on January 1, 2003. Since that time, the response from participating nations has varied. South Africa, for instance, has for years attempted to create an effective monitoring system, but disputes with major diamond traders such as De Beers have arisen over where diamonds will be made into parcels for export. In July 2004, UN inspectors concluded that the Republic of Congo was exporting diamonds at a rate about 100 times greater than its estimated production. The conclusion was that most of the diamonds being exported by the country had been smuggled in illegally in violation of the Kimberley Process. The Republic of Congo was suspended from the scheme, and its diamonds were barred from legitimate international markets.

In the United States, presidential executive orders have declared a national emergency to restrict the importation of rough diamonds into the country from Sierra Leone, followed by similar restrictions on diamonds from Liberia. Congress also responded with the 2003 Clean Diamond Trade Act, which requires annual reviews of the standards, practices, and procedures of any entity in the United States that issues Kimberley Process certificates for the export of rough diamonds.

Nongovernmental organizations have also been stakeholders in this conflict. Amnesty International carried out a nationwide Day of Action on September 18, 2004, to determine what steps communities were taking to keep conflict diamonds out of stores. Surveys of retailers led the group to conclude that the diamond industry was still failing to deliver on the promises that had been made by the World Diamond Council. The group also determined that there is a lack of consumer awareness about conflict diamonds: 83 percent of the jewelers who responded said that their customers rarely or never inquired about the source of the diamonds, often used as an excuse for not having a policy in place. The survey also found that most retail salespersons could not provide sufficient assurances to the public that their diamonds were conflict-free, in the form of written policies and warranties. Another NGO, Global Witness, estimated in 2002 that the terrorist group Al-Qaeda has laundered $20 million using conflict diamonds (Amnesty International 2006). As a result, the issue has grown from being an African problem to one that is tied to some of the most important issues for domestic security today.

Precious metals are also a major source of natural resource conflicts, although the disputes have been going on for a much longer time. Gold deposits that had been discovered in the 1980s or earlier were placed on hold in the late 1990s, when gold prices dropped to a twenty-year low of $350 per ounce or less, after banks decided that they no longer needed to keep such large stockpiles. By 2000, exploration and production had flattened out and then went into a decline. But by 2006 gold prices were on the rise again, approaching $550 per ounce and encouraging mining companies to look to previously untapped sources in Ghana, Russia, Australia, and Latin America. From 2002 to 2005, the exploration budgets for the world's gold mining companies tripled (Shore 2006).

The natural resources element of these conflicts is complicated by a concern about the environmental degradation caused by gold-mining operations and about the economics of war and human rights. World Bank economist Paul Collier examined forty-seven civil wars from 1960 to 1999 and found that nations earning at least a quarter of their yearly gross domestic product (GDP) from the export of unprocessed commodities face a far higher likelihood of civil wars than countries whose economies are more diversified (Collier 2001).

The No Dirty Gold Campaign is an example of how groups are placing pressure on retailers by getting companies to agree to the "Golden Rules." In order to ensure that gold is mined responsibly, retailers must identify and disclose the source of the gold they sell to make sure that their products do not include gold mined at the expense of communities, workers, and the environment. The campaign also calls upon consumers to sign the No Dirty Gold pledge so that mining companies will provide safe working conditions, inform communities that may be affected by mining practices, show respect for basic human rights, cover the cost of closing down and cleaning up mining sites, and disclose the social and environmental effects of their projects (No Dirty Gold 2006).

The gold issue is different from the one involving conflict diamonds in the sense that it deals more directly with environmental degradation, even though both resources involve mining. Both problems still depend on building consumer awareness about the sources of natural resources, and the willingness of governments to regulate or otherwise control valuable minerals. Both disputes must be dealt with on the international level, and nongovernmental organizations are taking the initiative to make sure that the problems are addressed as quickly and comprehensively as possible.

# Oil and Natural Gas Development

"Black gold" is as valuable in many countries as diamonds or other precious metals, especially industrialized nations that rely upon gas and petroleum products to fuel their economies. Rising demand and uncertain supplies have led to prolonged military conflict, even undeclared war, among countries whose thirst for oil and gas outstrips their own resources. As more developing countries become industrialized, the finite supply of these natural resources makes them even more valuable, and for some, even more worth fighting for. Unlike the more limited geographical availability of gold and diamonds, oil and gas are found on every continent except Antarctica, and even there, some experts believe that there are potential reserves. The natural gas industry operates in sixty-seven countries; the World Energy Council has members in more than ninety nations.

There are two primary problems that have led to conflicts worldwide: the environmental impact of exploration, drilling, and transporting oil and gas, and the effects of foreign governments and companies on a country's political, economic, and social structure—what researchers call the "human dimension" of environmental policy, especially in developing countries dependent upon petroleum and gas products for economic survival. These are not new conflicts; oil was discovered in the Middle East in 1908, and battles over control have been going on ever since.

Environmental degradation can take many forms. There is a considerable amount of research showing that the effects of oil on marine life from offshore exploration and drilling, or on leaks from damaged transport vessels, can be devastating, especially in the short term. Oil can contaminate marine life and smother it, or its toxic components and bioaccumulation can taint both animals and plants. Cleanup efforts can also damage the environment when certain types of chemicals are used, or surrounding reefs and the ocean floor are disturbed. On land, runoff from pipelines and wells can pollute groundwater and surface water, can pose a hazard through exposure, and can contaminate soil. The use of dynamite to build roads or as part of seismic surveys, burning forest lands to open up drilling facilities, or even human encroachment into previously undeveloped areas can injure or kill wildlife and disturb habitats. Gas flaring can result in deposits of soot on the ground and on structures, and has been associated with reduced crop yields.

Existing protective regulations and laws may not be sufficient to prevent the environmental damage caused by oil exploration and drilling in developing countries. In Ecuador, for instance, the government owns all of the natural resources and receives the benefits and profits of investment directly. A $1.3 billion oil pipeline built by a consortium of seven international corporations was routed through protected areas, including national parks and wildlife reserves. Other oil projects have been approved by the government on millions of hectares of forest that are the ancestral territories of a half-dozen indigenous communities. According to Amnesty International, some of the communities have vowed that they will never permit oil drilling on their lands, while others have been intimidated into believing that they have no ability to resist government troops. The government has also approved plans by Brazil's state-owned oil company to begin drilling in Yasuni National Park, even though the United Nations declared the area a protected biosphere preserve in 1989. Other U.S. and foreign-owned companies have also gained oil concessions in the national park.

While the environmental damage may be visible, it is more difficult to quantify or otherwise determine how drilling and exploration affect people and their way of life. In Ecuador, for instance, an estimated 50 percent of the country's budget is funded by oil exploration and production, so that any disruption can severely affect the country's oil revenues. Groups such as the Rainforest Action Network have documented increases in toxic contamination that have eventually led to disease and increased health risks from exposure. In Nigeria, where vast oil reserves were discovered during the 1950s, indigenous peoples in the oil-rich delta region have been driven from their homes or had their farms and crops ruined by pollution. The government has taken little action against multinational corporations that have failed to conduct cleanup operations, and those who have protested have been threatened or even killed. Above-ground pipelines are often subject to sabotage or rupture, and local authorities may lack the expertise to stop leaks and the pooling of oil, creating an additional danger.

In many regions where oil and gas are found, local residents derive little or no economic benefit from extraction. In the Caspian Sea region, the breakup of the former Soviet Union has allowed Western companies, previously shut out of the area during the communist regime, to make overtures to newly emerging governments. The discovery of significant deposits in the 1990s brought an influx of foreign investment, little of which has made its way into the lives of those living in Azerbaijan or Kazakhstan. The lack of regional cooperation, disputes over the legal status of the Caspian Sea, and weak or nonexistent environmental laws and regulations exacerbate the problem. Although some environmental laws have been enacted in Azerbaijan, Kazakhstan, Russia, and Turkmenistan, corruption is rampant, and regulation of oil and gas production is often considered a threat to the economy. Overlapping jurisdictions, the lack of regulations to implement statutes, and the uncertainty over the role of local and national authority continue to make the situation tense (U.S. Energy Information Agency 2003).

Chad "signifies the worst of Africa," according to the International Crisis Group in Pretoria, South Africa. Many people are aware of the humanitarian crisis in Darfur, in neighboring Sudan, but few know that the country has been politically unstable since independence in 1960. Much of the insurgency against the government is related to the nation's huge wealth in oil. Although it is the world's fifth poorest country, Chad has recently discovered oil fields where fuel can be transported 650 miles through a pipeline to Cameroon on the African coast. The pipeline was partially funded by the World Bank in a deal that was designed to create a model of how natural resources could fund programs for the poor. But in December 2005, Chad's president altered the poverty-reduction laws that were a part of the agreement, so the World Bank froze $124 million in loans earmarked for the country. In turn, the president threatened to suspend oil exports at a time in 2006 when the price was reaching record highs. The U.S. State Department, concerned because Africa could supply as much as one-fourth of U.S. oil imports by 2015, sent a negotiator to try to end the stalemate (McLaughlin and Soares 2006).

The promise of oil and money have led to violence and terrorism similar to that found in diamond and gold mining regions. Marxist guerillas have bombed oil facilities and kidnapped oil workers in Colombia, interrupting production and leading to price hikes. Drug cartels and paramilitary groups have seized upon oil and gas operations as sources of cash. Petroleum exploration sometimes clashes with territory claimed by rebel groups or for coca crops. In other politically volatile areas, such as central Asia, government officials are unable to guarantee worker safety or the protection of oil pipelines necessary to get products to production facilities. Proposed pipeline routes must cross over hundreds of miles of disputed territory, subject to sabotage or other types of damage.

Sovereignty issues also play a role in political unrest when oil and gas revenues are at stake. In the Spratly Islands, for example, the Philippines made their first claim to the area in 1975. Malaysia made several claims of its own in 1979, claiming that the islands are part of its continental shelf under the Law of the Sea, a claim also made by Brunei. China, which has maintained troops on one of the islands since 1956, asserts that Chinese navigators discovered the islands and that they have been under their control since the fifteenth century. Vietnam claims that the territory is part of the empire of Annam, having been so since the nineteenth century. Each country has its own name for the region's islands, and from time to time there have been military skirmishes over ownership. None of this would be important were it not for the fact that oil exploration and drilling began in the mid-1970s.

Iraq represents another type of sovereignty conflict over oil resources. Western oil companies have battled for the right to drill in Iraq since the early twentieth century, with increased demands after World War I. Oil drilling and production facilities were nationalized in 1972, leading to the exclusion of oil companies from the United States and United Kingdom, although the Iraq National Oil Company gave large contracts to France, Russia, and China. In 2003 the United States invaded and overthrew the government of Saddam Hussein, assuming control over the country as the Coalition Provisional Authority. Occupying forces have faced armed resistance ever since, and violence has spilled over to include Iraqi police and citizens as the country deals with civil strife. Insurgents have repeatedly sabotaged or destroyed the country's infrastructure. The United Nations assumed a limited role in August 2004 in an attempt to smooth the political transition and to assist in the reconstruction of the country, which is burdened by debt. What Iraq does have, however, is the world's second largest proven reserves of high-quality oil.

Iraq had been subject to economic sanctions by the UN Security Council until after the U.S. invasion, when a May 2003 resolution called for a development fund for the country. The idea was to administer proceeds from the export sales of Iraqi oil, along with other assets seized from the former regime, to help rebuild the country. After an oil-for-food scandal that rocked the United Nations, the Iraqi people appear to have been cut out of many of the decisions about their natural resources. The removal of government subsidies, combined with the ongoing U.S. occupation, has created disputes between foreign oil countries seeking greater access to the valuable reserves and the country's oil workers, who are attempting to rebuild the industry without foreign interference.

The newly written 2005 constitution gave the United States a major role in production-sharing agreements, even though the Iraqi oil workers' union opposes denationalization. In 2006 political change made the future of the oil fields—and who governs their operation—increasingly uncertain. Within the semiautonomous Kurdish parliament, for instance, leaders have begun making plans for their own Ministry of Natural Resources to control oil and natural gas in three northern provinces, where a Norwegian company has already received a contract to survey and drill for oil. At issue is the fact that the Kurdish agreement was made without the involvement of Iraq's central government, weakening its power (Global Policy Forum 2006).

Another dimension of oil and gas exploration is that of human rights, especially in Latin America. Texaco began its exploration for oil in Ecuador in 1964, when environmental concerns were just beginning to be voiced in the United States and the major issues were water and air pollution and wilderness preservation. The company entered into a partnership with Petroecuador to conduct drilling operations, and one study alleges that between 1972 and 1992 more than 19 billion gallons of toxic wastewater were intentionally dumped into the region, with 16.8 million gallons of crude oil being spilled into the forest. The result, experts believe, is an increase in the number of local people suffering from health problems, including higher cancer rates and higher rates of spontaneous abortion. Texaco (which became ChevronTexaco after merging with Chevron Oil in 2001) has denied any correlation between public health hazards and its drilling operations, and it has refused to offer compensation to affected communities. A lawsuit was filed against the company in 1993 on behalf of 30,000 Amazon residents affected by the pollution and oil spill, and the matter is still mired in the courts in both countries (Amnesty International USA 2006).

One of the lead organizations in the dispute is Amnesty International, which has urged ChevronTexaco shareholders to protest the company's actions by calling for a report on new initiatives that address the health and safety concerns of communities affected by the contamination in Ecuador. Groups such as Trillium Management (a socially responsible investment firm) and the New York State Common Retirement Fund, which has invested heavily in the company, have cofiled resolutions asking for more information for its investors at the firm's annual shareholder meetings. Another group, Amazon Watch, has sent investigative teams to Ecuador, including representatives from shareholder institutions, to determine claims against the company. While in the country the representatives visited waste sites and interviewed residents who described how the oil operations had polluted local waterways, causing a 70 percent decline in agricultural productivity and increasing rates of cancer, birth defects, respiratory infections, and unclassifiable mysterious deaths (ibid.).

Some environmental organizations have found that it is possible to work with international companies and state-run industrial developers. In Venezuela, for example, the government's oil company, Petroleos de Venezuela, is prospecting in an area of about 64,000 square miles off the Caribbean coast. Inasmuch as oil and gas sales account for about a third of the country's gross domestic product, there is great interest in developing offshore sites to bolster the struggling economy. The area is also ecologically sensitive, as the habitats for endangered sea turtles, dolphins, and manatees, and its coral reefs, are among the healthiest in the world. The government sought the Nature Conservancy's assistance in identifying sensitive sites and ways to minimize potential damage from drilling operations. Along with the Institute of Technology and Marine Sciences at Simon Bolivar University, the organization is sharing technology such as satellite imagery and aerial photographs to identify areas potentially susceptible to disturbance. The partnership is an unusual one, but it could serve as a model for other oil and gas development (Butvill 2005).

But some conflicts defy resolution. In Nigeria, for example, militants have repeatedly sabotaged and attacked oil and gas pipelines belonging to Royal Dutch Shell, Africa's leading oil exporter and the fifth largest supplier to the United States. Attacks on facilities and workers in the Niger delta in early 2006 caused an increase in world oil prices after the disruptions became more common, even though U.S. stockpiles had been slowly increasing and energy needs traditionally drop in the middle of the year. Insurgents within Nigeria vowed to cut the output of oil by 30 percent by March 2006 as part of their planned violence leading to national elections in 2007.

Civil strife like this, especially when oil and gas supplies are insecure in other parts of the world, have led to questions about whether prices will continue to increase to a point where consumers have few choices for their energy sources. Both the United States and the United Nations have been reluctant to interfere when military resources are already stretched thin in the Middle East, the major source of imported oil.

# Timber and Forest Destruction

About 8,000 years ago, forests covered some 40 percent of the earth's landmass, or an estimated 6 billion hectares. Ever since humans discovered fire, forested areas on the planet have been burned or cut down as sources of fuel, building materials, road materials, paper products, or to clear land for agriculture. Human migration, and the accompanying clearing of forest areas, affected the Middle East, the Mediterranean watershed, South Asia, and the Far East for the next 7,500 years. About 500 years ago forests in North America began to disappear, as did areas in coastal Brazil and in the Caribbean, where sugar plantations took over from the trees. In Europe, the Industrial Revolution reduced forested areas as wood was used for furnaces and factories, and by the late 1800s ancient forests were virtually gone. Between 1850 and 1980, an estimated 15 percent of the world's forests and wooded areas were cleared (Chew 2001; Roper and Roberts 1999).

Modern assessments of forest destruction are usually based on satellite mapping and imagery that classifies the landscaped area. The UN Food and Agriculture Organization (FAO) has conducted one of the most comprehensive assessments of forests every five to ten years since 1946. In its 2005 report the agency collected and analyzed data from 229 countries and territories, examining about forty variables including the extent of forest resources, forest health, biodiversity, the productive and protective functions of forest resources, and their socioeconomic functions. Both undisturbed primary forests and managed ones (including plantation forests) were studied to provide information on world-wide forest conditions.

Primary forests are those composed of native species where there is no clearly visible indication of human activities and where the ecological processes have not been significantly disturbed—about one-third of the world's forested area. Some forests that have been disturbed can return to a primary forest state over time if there is no human intervention, as has been the case in several European countries and Japan. Plantation forests are defined as those that consist primarily of introduced species; they make up about 140 million hectares, or an estimated 3.8 percent of the world's forested areas. These forests are used primarily for wood and fiber production, although about one-fifth of the plantations are established for soil and water conservation. Plantation forests have been increasing at the rate of about 2.8 million hectares per year from 2000 through 2005 (UN Food and Agriculture Organization 2006).

The study notes that the world's total forested area is just under 4 billion hectares, but those lands are unevenly distributed. For example, the FAO found that sixty-four countries with a combined population of 2 billion (about one-third of the earth's population) have less than 0.1 hectares of forest per capita. The ten most forest-rich countries, in contrast, account for two-thirds of the world's total forest area. Seven countries or territories have no forest at all, with fifty-seven other countries having forest on less than 10 percent of their total land area (ibid.).

About 13 million hectares of forest are lost each year, although efforts such as replanting have lowered the net loss to about 7.3 million hectares each year between 2000 and 2005. That figure compares with a net loss of 8.9 million hectares per year between 1990 and 2000, in part because of increased efforts at restoration and natural landscape expansion. The greatest forest loss is occurring in Africa and Latin America, while Europe's forests are expanding and Asia's now show evidence of a slight net gain (ibid.). However, forestry experts admit that it is nearly impossible to get a precise number of hectares lost each year, because of difficulties in getting information from some developing countries when land is logged illegally or because of widespread forest fires.

In Brazil, for instance, the government announced in 2005 that the rates of deforestation had dropped by as much as 50 percent because of new policies. But researchers from Stanford University believe that the figure may be inflated, and that the Brazilian Amazon is being logged at more than twice the rate previously thought. Their study, which included satellite images of areas that had been selectively logged as well as clear cut, showed that in the five states that make up about 90 percent of all deforestation in the region, selective logging added 60 to 123 percent more damage to forest areas than reported previously (Mastny 2006).

Logging is one of the most visible and often damaging forest practices, especially when all of the trees in a specific area are cleared. About one-third of the world's forests are used for wood production, amounting to about 3.1 billion cubic meters in 2005. Timber harvests are decreasing in Asia but increasing in Africa, with about one-half of the removed wood used for fuel. Rogue logging continues in countries such as Cambodia, Ecuador, Indonesia, and Liberia.

In Canada, intensive logging in temperate rain forests has been the focus of numerous controversies, especially in British Columbia. While timber harvesting is anathema to many environmental groups, it is especially contentious in areas where the sovereignty of indigenous peoples is concerned. The Great Bear Rainforest Campaign, working in conjunction with the leaders of a dozen First Nations, continues its efforts to stop or defer logging in the Mid-Coast region of the province. Organizations such as the Forest Action Network and Greenpeace Canada have conducted direct action protests, including tree sits and blockades modeled on activism in the forests of the U.S. Pacific Northwest. The Canadian "War in the Woods" has been going on since the mid-1990s in an effort to protect 1,000-year-old cedar trees and ancient spruce, as well as habitat for grizzly and black bears.

In February 2006, the British Columbian government agreed to provide strict protection for one-third of the Great Bear Rainforest (an area of about 2 million hectares) and to require the logging industry to implement an ecosystem-based management system for the other two-thirds of the rain forest by 2009. Environmental leaders counted the agreement as a victory, saying that "common sense has prevailed. Today we celebrate a magnificent victory with the thousands of activists who stood their ground in British Columbia, every cyberactivist who ever sent an e-mail action alert on this issue, and the millions of people worldwide who have let it be known through their votes and their consumer choices that the world's remaining ancient forests need to be preserved" (Greenpeace 2006).

Forest cover is also destroyed by natural disasters such as floods, hurricanes, fires, drought, and insect infestations, which result in the loss of an estimated 104 million hectares of forest each year. In 1997 and 1998, for instance, forest fires in Southeast Asia, especially in Indonesia, burned an estimated 2 million hectares.

The impacts of timber and forest destruction have varying implications for humans and for biodiversity. There are tens of millions of species on earth, and only about 1.5 million have been discovered and named, with an estimated 137 species going extinct every year. On a positive note, however, considerable efforts are being focused on increasing the amount of land classified as protected areas as a way of slowing species loss. The FAO study found that 11 percent of the world's forests are now designated for the conservation of biological diversity, having increased by an estimated 96 million hectares since 1990.

Once the land has been cleared for agriculture or grazing, farmers often use substantial amounts of fertilizer or pesticide, and the runoff from the chemicals seeps into the soil and water below. These substances weaken the ecosystem, making it difficult for trees to regrow. Gold mining in parts of Africa has a similar effect on tree growth.

There is a human dimension to illegal logging as well. In February 2006, an estimated 3,000 people died in a devastating landslide in Saint Bernard town in the Leyte province of the Philippines. Representatives of the environmental group Greenpeace echoed residents' statements that illegal logging operations that started in the 1970s above small villages had caused the rain-soaked ground to give way. The area had received 27 inches of rain—twice the normal amount—and past years' chainsaw logging on Mount Guinsaugon had weakened the soil. Although logging is illegal, Greenpeace said that a combination of poor governance and corruption has hampered enforcement of the law. The combination of logging and extreme weather events caused by climate change "should be taken seriously by the Philippine government," said a Greenpeace spokesman. "The scale and frequency of similar tragedies in the past should have, long before, already provoked the government into action to address the seemingly perennial problems of floods and land-slides at the source" (ABS-CBN News 2006).

Trade in illegal wood products creates a different kind of problem. The United Kingdom is believed to be the largest importer of illegal tropical timber in the world, undermining efforts at protecting forests and leading to criminal activity.

Another problem associated with deforestation is the impact on the global carbon cycle. When a forest is cut and burned to make room for agriculture and grazing, for instance, the carbon that was stored in the trunks of the trees joins with oxygen, released into the atmosphere as carbon dioxide. The National Aeronautics and Space Administration (NASA) estimates that from 1850 to 1990, worldwide deforestation released 122 billion metric tons of carbon into the atmosphere. Currently, about 1.6 billion metric tons are being released each year, enhancing the greenhouse effect that may contribute to an increase in global warming (National Aeronautics and Space Administration 2006). Forests become carbon sinks, sequestering carbon by storing it in litter and soil as well as in the trees themselves.

Sustainable forestry is not a one-size-fits-all solution. An alternative practice to clear-cutting, for example, is selective logging, which requires marking and cutting specific trees, rather than a whole area at one time. One study in Indonesia found that when only 3 percent of the trees were cut, a logging operation would actually damage nearly half of the surrounding forest. But in tropical forests, selective logging still provides seeds for reforestation and shade for trees beginning to grow (ibid.).

Another management technique involves reduced-impact logging, or RIL, which can be 10 to 15 percent less expensive than conventional logging methods. RIL involves an array of harvesting techniques that reduce damage from road building and skid-trails, ensures faster recovery and better survival of residual trees, shortens cutting cycles, and helps to establish inventory control. According to the Tropical Forest Foundation (TFF), RIL techniques can be adapted to fit individual biophysical systems and economic conditions. The methods include careful tree felling and machine use that increase worker safety and reduce wood waste by more than 60 percent (Holmes et al. 2006).

Because tropical forests are often found in developing countries, it is often difficult to resolve disputes over the appropriate use and protection of a nation's trees. Western environmental organizations, for instance, have sought protection for tropical rain forests. But key to any solution is finding alternative sources of income for impoverished populations seeking to improve their livelihoods, or those relying upon forest resources. In Liberia, where the UN Security Council placed a sanction on timber exports, the halt in trade is said to have harmed ordinary citizens who are dependent upon forests for economic survival.

Shade agriculture refers to the practice of not cutting the original trees in the rain forest, providing shade for crops such as coffee or chocolate. If a farm is abandoned and the crops are no longer being harvested, much of the forest grows back very quickly, sometimes within twenty years.

Protected area networks represent one form of direct action to reduce deforestation, but funding is limited for long-term efforts, especially research and training. Unless field staff are paid reasonably well, the opportunity for corruption and illegal logging is always present.

# Water Resources and Scarcity

In 2006, a UN special investigator warned that 11 million people in East Africa are faced with starvation because of a combination of drought and conflict. In Djibouti, Ethiopia, Kenya, Somalia, and Tanzania, prefamine conditions were increasingly being reported, and the World Meteorological Union warned that substantial rain was unlikely. The drought, reported to be the worst in decades, along with poor governmental planning, exemplifies the problems faced in many developing countries. Aid organizations such as Oxfam are unable to provide more than a token supply to remote areas, forcing families to make the choice between allowing their animals, which are their life savings, to starve or to die of thirst. Temperatures of over 100 degrees are common, and the cracked earth is testament to how long it has been since the last rains. Although some water is available from private vendors, the prices have soared from about 3 cents for a 20-liter jug in normal times to about a dollar, more than most people earn in a day when work is available (Lacey 2006).

Water scarcity is more than just a question of availability. It is a deeply divisive international issue that some, including the vice president of the World Bank, have warned will be the source of the world's next deadly wars. Whether those projected conflicts take place among military forces in the Middle East, or among lines of refugees in Somalia hoping for a few liters of water that must last a week, scarcity of this vital resource is pitting the have nots against the have mores.

Water scarcity can be defined in numerous ways, depending upon one's needs. Generally, the term is applied to situations in which the annual availability of water is less than 1,000 cubic meters per person. To put this in context, in the Middle East, where water scarcity is considered serious, Israel has available less than 300 cubic meters, Jordan around 100 cubic meters, and in Gaza and the West Bank, well below 100 cubic meters of renewable water per person each year (Worldwatch 2006). In terms of use, it takes about 1,000 cubic meters to grow a ton of grain. In terms of production, a proposed seawater desalination plant in Algiers can produce 200,000 cubic meters of potable water a day, or about 53 million U.S. gallons. The world's largest water reclamation plant, the Sulaibiya facility in Kuwait, purifies municipal wastewater, producing about 100 million gallons per day, mostly for industrial and agricultural use.

Most of the world's drinking water can be traced to groundwater, such as rivers and streams, which flow into aquifers beneath the surface of the earth. According to one study, there are 263 international river basins, covering 45.3 percent of the earth's land surface, hosting about 40 percent of the world's population and accounting for about 60 percent of global river flow. Territory in 145 nations falls within international basins, and thirty-three countries are located almost entirely within those basins; seventeen nations, for instance, share the Danube River basin (Wolf et al. 2005). Control of rivers and the basins where they flow is one part of the reason for the differences in water availability from one region to another.

Statistics on the magnitude of the problem vary, but they all point to many developing countries that are in dire straits because of water scarcity. Some estimates indicate that by 2015, some 3 billion people, or about 40 percent of the world's population, will find it difficult or impossible to mobilize enough water to satisfy their food, industrial, and domestic needs. In addition, there are spillover effects leading to poverty and refugee migration that place additional pressures on other nations and regions. There is a sense, even among political leaders who have been told otherwise, that water is an abundant and renewable natural resource. Few have recognized that the problem stems in large part from overconsumption as the world's population grows, as does demand for agricultural products and growing industrialization. Even more serious in some regions is the continuing deterioration of drinking water quality, which often leads to disease. Current international plans to address the issue are unlikely to be met by the goal year of 2015 (Biswas 2005).

Another problem relates to what experts call problem displacement. Countries such as Israel, where water is scarce, divert their few liquid resources to cities and industries, where the "value" per gallon is higher than it is for agricultural production, which requires considerably more water at a greater "cost." The nation imports agricultural goods, which only transfers the problem to the exporting countries, where water may be just as scarce (Princen 2003).

In northern Africa, several types of water crises are interconnected. Drinking water is often in short supply, rainfall is limited and unpredictable, and the existing infrastructure leads to evaporation and immense losses of water. The residents in some urban areas receive water only once every three days. The United Nations estimates that, in other parts of the continent, patterns of unsustainable use of water, poor management, pollution, increasing consumption, and rapid population growth are responsible for numerous conflicts. The UN estimates that by 2025, one out of two Africans will be living in countries facing water scarcity.

Researchers at the Pacific Institute for Studies in Development, Environment, and Security have developed a chronology of conflicts over water that dates back to 2500 B.C. and the border wars between the city-states of Lagash and Umma. The timeline identifies dozens of disputes from biblical times (1200 B.C., and the accounts of Moses parting the Red Sea to save Jews trapped by the Egyptian army) and multiple wars in Babylon to the early history of the United States. In most of these instances the military (or invading forces) used water strategically, destroying dams, diverting rivers, flooding towns, creating liquid barricades, or dynamiting canals. The control of water resources that is the foundation of contemporary disputes over scarcity is traced back to 1898, when Egypt, France, and Great Britain battled over the headwaters of the Nile River, and to 1907, when Los Angeles sought to divert water from the Owens Valley to satisfy the needs of thirsty Southern Californians (Gleick 2004).

More recently, control of water resources has resulted in partitions dividing the Ganges River between Bangladesh and India

(1947); the partition of the Indus basin between India and Pakistan (1947); disputed territory between Sudan and Egypt (1958); disputed territory between Israel and Syria in 1962, in 1964, and again in 1965–1966; disputed territory between Brazil and Paraguay from 1962 to 1967; disputes among Argentina, Brazil, and Paraguay in the 1970s over the Parana River; disputes between Angola and South Africa in 1975; tensions over the Blue Nile between Egypt and Ethiopia since 1978; disputes between Lesotho and South Africa in 1986; since 1991, disputes between Karnataka and Tamil Nadu (India); problems over access to water by Namibia, Botswana, and Zambia in 1999–2000; destruction of water supplies in the West Bank settlement of Yitzhar by Palestinian force in 2001; and, in 2001, disputes in Macedonia (ibid.). Most of the disputes that have occurred since World War II are ongoing, punctuated by only brief periods of calm before strife reoccurs.

The Worldwatch Institute estimates that, since World War II, more than 45,000 dams have been built, and they generate about one-fifth of the world's electricity. The water from those projects also provides irrigation water for agriculture that produces about 10 percent of the world's food (ibid.). The conflict over dams and water scarcity arises when the facilities are built near the ancestral homes of indigenous people. Rural and ethnic minorities are sometimes the interests that "lose" because dam building often calls for such groups to be moved or evicted from land that has been occupied for centuries. Resettlement may be poorly planned, or may result in minimal if any compensation for those forced to relocate. Cultural considerations are sometimes ignored, and religious or spiritual sites might be destroyed behind reservoirs. Informed prior consent, while desirable, may not result in acceptance.

Another form of conflict arises when one project affects communities in another area. A good example of this can be found in Mexico, where the Federal Electricity Commission is planning to build a $1 billion hydroelectric dam on the Papagayo River. Officials say that the dam is needed to supply the tourist resort of Acapulco, twenty miles away, and that the project would provide water for the next fifty years. Acapulco is the top source of income, and its population is expected nearly to double over the next fifteen years. A second project, in the western state of Jalisco, was planned but stalled because of protests.

The Papagayo project is being opposed by farmers who say that the dam will cause water shortages that will dry up their farms and fisheries. The government admits that water will be scarce for one to two years as the dam fills, but they have promised to guarantee a minimum flow in the interim. Locals say that water alone is not the issue. The dam would ruin a lifestyle of farming and fishing that has existed for generations. Others support the project because it would bring higher-paying jobs to the region. The conflict has become more heated because the group has been advised by radical opposition movements. In opposing another governmental project in 2002, radical leaders kidnapped officials, tied them to a tanker truck, and threatened to blow it up. In the region where the proposed dam would be built, hundreds of machete-wielding farmers briefly seized a pumping plant, cutting off water to Acapulco just before the resort opened for the major spring tourist season. The protestors ended a two-day blockade when the government provided them with compensation for some land and the release of one of their colleagues.

A similar dispute involves the proposed Siberia–Central Asia Canal, which would divert Siberian rivers into the desert along a 2,225-kilometer corridor. The proposed dam was designed as a source of valuable income for Russia, which had planned to sell the water from the Ob and Irtysh rivers (less than 10 percent of the flow would be diverted) at a cost between $12 and $20 billion. The excess water would be sold to downstream countries where water is scarce by diverting the canal's water into the Amu Darya and Syr Darya rivers to serve the five Central Asian states of Krygyzstan, Tajikistan, Kazakhstan, Uzbekistan, and Turkmenistan. These nations rely upon water for irrigation of thirsty crops such as cotton and rice; agricultural use and diversion are already drying up the Aral Sea, which UN experts predict may not exist by 2020.

Russia's plans, which have been called exorbitantly expensive and not feasible, would help its economy, but the impact of the canal on the five beneficiary countries would potentially lead to intense conflict in Central Asia. Competition is fierce for the already scarce local sources of water, about half of which are found in Tajikistan. That country's leaders admit that they do not have the resources to maintain, repair, and operate its current irrigation infrastructure, and many citizens still do not have access to clean drinking water. The region's use of water for irrigation is inefficient, since water is lost through evaporation and in its poorly built canals. The Russian canal project looks promising, at least to some (Blagov 2003).

Although conflicts relating to water resources have been around for thousands of years, the global news media have seized upon the issue by graphically illustrating the magnitude of the problem in the regions affected, and the people who are dying because they have no access to water. Some researchers, however, argue that the apocalyptic warnings are unsubstantiated, since no nations have specifically gone to war over water resources for thousands of years. Instances of cooperation outnumbered conflicts by more than two to one between 1945 and 1999, according to the Worldwatch Institute. Thirty of the thirty-seven disputes within that period involved Israel and one of its neighbors. Non-Mideast cases accounted for only five events, while during the same period, 157 treaties were negotiated and signed. "Because water is so important, nations cannot afford to fight over it. Instead water fuels greater interdependence. By coming together to jointly manage their shared water resources, countries build trust and prevent conflict" (Wolf et al. 2005).

Low-level water cooperation has occurred in some areas of the Middle East, however, despite the extent of the bitterness and hostility among affected nations. The United Nations has worked since the early 1950s to build agreements between Israel and Jordan, even when the two countries were officially at war. When a peace agreement was reached in 1994, there were already standards and norms that had been informally reached years before, providing a basis for trust.

The resolution of water conflicts has been hampered by the lack of good governance—specifically, by corruption within national and local governments. But there are several strategies that have been recommended to help end some of the disputes, to bring about peaceful interaction among the affected parties. The Worldwatch Institute has developed several ideas for future initiatives:

- Identify and utilize more experienced facilitators who are perceived as truly neutral. Researchers cite the example of the World Bank's efforts to facilitate the Nile Basin Initiative as a model for the types of skills that could be used in other river basins.
- Be willing to support a long process that might not produce quick or easily measurable results. In some cases, nations have had to make a decades-long commitment to a project because of the length of a

project cycle. That has been the case with Sweden's agreement to commit resources to Africa's Great Lakes for twenty years.

- Ensure that the riparians themselves drive the process. Some of the failures in past conflict-resolution attempts have been a result of negotiators perceived as outsiders who have little stake in the outcome. Worldwatch recommends that negotiations emphasize the role of local interests, and that efforts be made to increase the capacity of excluded, marginalized, or weaker groups to ensure that their voices are heard.
- Strengthen water resource management. There are many water resource institutions already operating at the local or regional level, but many of them lack the necessary capability to analyze data, encourage public participation, or develop long-term management plans. The expertise of indigenous people can be combined with modern technology and the knowledge of nongovernmental organizations that have already built relationships and could assist in capacity building.
- Balance the benefits of closed-door, high-level negotiations with the benefits of including all stakeholders—NGOs, farmers, indigenous groups—throughout the process. This step involves explicitly inviting and encouraging the participation of all relevant parties in the dispute before any decisions are made. Sometimes these groups will buy into a project that they might otherwise reject because they have had an opportunity to participate in decision-making (ibid.).

The ultimate goal of any of these strategies is to avoid conflict. Water resources are valuable, in some regions more so than gold or precious minerals. But there is also evidence that the water wars of the past are different today, and that new initiatives will be needed to make sure that scarcity issues remain on the negotiating table rather than being exported to the killing fields.

### References
ABS-CBN News. 2006. "Greenpeace: Logging, Climate Change Aggravated S. Leyte Mudslide." Website at www.abs-cbnnews.com (accessed February 18, 2006).

Amnesty International USA. "Conflict Diamonds Update: Kimberley Process and Industry Self-Regulation" at www.amnestyusa.org/diamonds (accessed April 15, 2006).

Amnesty International USA. 2006. "Ecuador: Oil Rights or Human Rights." Website at www.amnestyusa.org (accessed April 17, 2006).

Biswas, Asit K. 2005. "An Assessment of Future Global Water Issues." *International Journal of Water Resources Development* 21, no. 2 (June): 229–237.

Blagov, Sergei. 2003. "Parched Central Asia Looks to Russia." *Asia Times Online* at www.atimes.com (accessed April 1, 2006).

Butvill, Dave Brian. 2005. "When Oil and Conservation Mix." *Nature Conservancy* 55, no. 4 (winter): 14.

Chew, Sing C. 2001. *World Ecological Degradation: Accumulation, Urbanization, and Deforestation, 3000 B.C.–A.D. 2000.* Walnut Creek, CA: AltaMira.

Collier, Paul. 2001. "Economic Causes of Civil Conflict and Their Implications for Policy." In *Turbulent Peace: The Challenges of Managing International Conflict,* edited by Chester A. Crocker, Fen Osler Hampson, and Pamela Aall. Washington, DC: United States Institute of Peace Press.

Gleick, Peter. 2004. "The World's Water: Water Conflict Chronology." Website at www.worldwater.org (accessed April 1, 2006).

Global Policy Forum. 2006. "Oil in Iraq." Website at www.globalpolicy.org (accessed April 23, 2006).

Greenpeace. 2006. "Great Bear Rainforest: Saved." Website at www.greenpeace.org/usa (accessed February 18, 2006).

Holmes, Thomas P., et al. 2006. *Financial Costs and Benefits of Reduced-Impact Logging in the Eastern Amazon.* Alexandria, VA: Tropical Forest Foundation website at www.tropicalforestfoudnation.org (accessed February 14, 2006).

Kennes, Erik. 2005. "Footnotes to the Mining Story." *Minerals and Energy* 20, no. 1 (March): 23–28.

Lacey, Marc. 2006. "Amid Deep Drought, Africans Are Dying." *Arizona Republic,* February 19, p. A28.

Masland,Tom, et al. 2000. "In Search of Hot Rocks." *Newsweek* 136, no. 2 (July 10): 30.

Mastny, Lisa. 2006. "New Imaging Techniques Reveal Greater Amazon Logging." *Worldwatch* 19, no. 1 (January–February): 5.

McLaughlin, Abraham, and Claire Soares. 2006. "Oil Wealth and Corruption at Play in Chad's Rebellion." *Christian Science Monitor* (April 21) at www.csmonitor.com (accessed April 22, 2006).

National Aeronautics and Space Administration. 2006. "Tropical Defor-estation." Website at www.earthobservatory.nasa.gov/Library/Deforestation (accessed February 14, 2006).

Natural Resources Canada. "Kimberley Process for Rough Diamonds–Background." Website at http://mmsd1.mms.nrcan.gc.ca/kimberleyprocess (accessed April 4, 2006).

No Dirty Gold website at www.nodirtygold.org (accessed October 18, 2006).

Orogun, Paul. 2004. "'Blood Diamonds' and Africa's Armed Conflicts in the Post–Cold War Era." *World Affairs* 166, no. 3 (winter): 151–161.

Perlez, Jane, et al. 2005. "Tangled Strands in Fight over Peru Gold Mine." *New York Times*, October 25, A1.

Princen, Thomas. 2003. "Principles for Sustainability: From Cooperation and Efficiency to Sufficiency." *Global Environmental Politics* 3, no. 1: 33–50.

Roper, John, and Ralph W. Roberts. 1999. *Deforestation: Tropical Forests in Decline*. Forestry Issue Paper (January) website at www.rcfa-cfan.org (accessed November 19, 2005).

Shore, Sandy. 2006. "Gold's High Prices Boost Production, Global Exploration." *Arizona Republic*, January 17, D2.

Stevenson, Mark. 2006. "Mexican Dam Project Faces Growing Farmer Opposition." *Arizona Republic*, April 7, A10.

Tamm, Ingrid J. 2004. "Dangerous Appetites: Human Rights Activism and Conflict Commodities." *Human Rights Quarterly* 26, no. 3: 687–704.

UN Food and Agriculture Organization. 2006. *Global Forest Resources Assessment 2005*. Rome: Food and Agriculture Organization.

U.S. Energy Information Agency. 2003. *Caspian Sea Region: Environmental Issues*. Website at www.eia.doe/gov (accessed April 17, 2006).

Wolf, Aaron T., et al. 2005. "Water Can Be a Pathway to Peace, Not War." *Worldwatch Global Security Brief* #5 (June) website at www.worldwatch.org (accessed April 1, 2006).

Worldwatch Institute. "Water Conflict and Security Cooperation." *State of the World 2005 Trends and Facts* website at www.worldwatch.org/features (accessed April 1, 2006).

# 4

# Chronology

**1626**      Plymouth Colony in Massachusetts regulates cutting and sale of timber on colony lands to control supply.

**1631**      Massachusetts Bay Colony passes ordinance providing penalties for setting destructive wood fires.

**1781**      Thirteen original colonies agree to cede the western reserve lands back to the government, creating the public domain lands.

**1785**      General Land Ordinance gives the federal government the right to grant public lands to new states to produce revenue, and initiation of a program of land grants for schools upon designation of statehood. Surveys conducted of the public lands were to conform to a rectilinear system based on lines of longitude, establishing a pattern of land ownership in 24-by-24-mile squares.

**1787**      Ordinance of 1787 provides that estates in the Northwest Territories could be bequeathed by wills and conveyed by lease, bargain, and sale as part of the transition to fee simple ownership.

**1803**      Louisiana Purchase doubles the size of the public domain by adding 530 million acres purchased from France for $15 million. The acquisition opens up the heartland of the continent and provides new trade routes along the Mississippi River.

1832    George Catlin, a lawyer who traveled throughout the frontier, writes of the benefits of preserving and protecting pristine wilderness, believed to be the first mention of wilderness preservation.

1832    Congress reserves four sections of the area around Hot Springs, Arkansas, for "future disposal" and sets the precedent for reserving lands for future designation as national parks.

1841    Log Cabin Bill permits any head of household, widow, or single man over age twenty-one, who is a citizen, to acquire up to 160 acres of surveyed, unoccupied, unreserved public land by settling upon it and paying $1.25 per acre, provided that the individual did not already own more than 320 acres within the United States.

1848    Gold is discovered in California; miners establish their own authority to create mining districts and property rights in absence of a state or federal presence.

1862    Homestead Act formalizes the idea of allowing settlers to claim sufficient public land to support a family, usually up to 160 acres per head of household.

1864    Congress enacts legislation granting the Yosemite Valley and Mariposa Big Tree Grove to the state of California to protect the area from haphazard tourism and land claims.

1865    Landscape architect Frederick Law Olmsted lays forth the philosophical foundation for the preservation of the Yosemite Valley in a report to California's governor, providing the framework for future national park policies.

1867    Purchase of Alaska marks the end of the period of acquisition of public lands as part of the public domain.

1869    Charles Cook, David Folsom, and William Peterson journey to Yellowstone and confirm the accounts of Jim Bridger, whose stories about the region were so grandiose that most officials and the public disbelieved them.

1869    John Wesley Powell completes the first successful transit of the Grand Canyon, giving him a national reputation for exploration and surveying of the West.

1870    Washburn-Langford-Doane expedition discovers the Upper Geyser Basin and Old Faithful geyser. Publicity from the exploration encourages public interest in Yellowstone, and Congress begins to consider establishing some type of protective status for the area.

1872    General Mining Law establishes U.S. mineral policy and regulations, including the rights of private citizens to explore and purchase public lands.

1872    Legislation designates 3,400 square miles in Yellowstone as the country's first national park as Congress calls for the "preservation, from injury and spoliation, of all timber, mineral deposits, natural curiosities or wonders, and their retention in their natural condition."

1873    Coal Lands Act requires development of a competitive leasing system, with bids based on fair market value. Coal is never included under the Mining Law of 1872.

1878    Enactment of the Free Timber Act allows the logging of timber within the public domain, whether for mining development, agriculture, or domestic purposes.

1878    Public Lands Commission is created; the agency recommends that all public lands be withdrawn from sale or disposal, but Congress disregards its recommendations.

1879        U.S. Geological Survey is established to investigate and assess mineral resources and administer mineral leases on public lands.

1879        John Wesley Powell's *Report on the Lands of the Arid Region of the United States* becomes the basis for the reform of the corrupt General Land Office, transfer of survey responsibilities to the Coast and Geodetic Survey, and the recommendation for the creation of communities along bodies of water.

1883        Congress authorizes the secretary of war, upon request of the secretary of the Interior, to provide army troops to prevent trespassers and intruders from entering Yellowstone National Park to hunt or to destroy objects.

1883        Division of Forestry is established by the secretary of agriculture.

1889        Congress appropriates funding to repair the damage at the Casa Grande ruins in Arizona, recognizing the importance of the archaeological site that was being vandalized and leading to the creation of the first national monument.

1890        Sagebrush Rebels introduce legislation to cede all unappropriated lakes and rivers to state or territorial control.

1890        Legislation establishes the nation's first national military park at Georgia's Chickamauga Battlefield and Tennessee's Chattanooga Park.

1891        Forest Reserve Act halts the sale of federal lands and allows the president to set aside forest reserves by executive order.

1891        Maine creates the first state forest commission with a fire suppression mission.

1892    Sierra Club is founded in San Francisco as the first major organization seeking to preserve wilderness areas.

1894    Yellowstone Game Protection Act (Lacey Act) provides additional protection for the park's wildlife and makes it a crime to hunt, kill, wound, or capture any bird or wild animal within the park's boundaries, one of the country's first environmental crimes.

1894    North Carolina Press Association petitions Congress for a national park, seeking to obtain protection for the Appalachian Mountains.

1897    Organic Act gives the Department of Interior authority to manage forest reserves to protect navigable waterways, prohibits clear-cutting, and requires the agency to secure a continuous supply of timber for public use.

1902    Newlands Act provides that all sales of arid or semi-arid public lands be deposited into a reclamation fund from which irrigation projects benefiting both public and private lands are to be financed. This becomes the source of funding for the Reclamation Service.

1904    Southern Pacific Railroad completes El Tovar Hotel in Grand Canyon National Park, further cementing relations between the railroads and tourism in the parks.

1905    U.S. Forest Service is created within the Department of Agriculture, with Gifford Pinchot as its first chief, with the mission of managing and protecting timber supplies in what becomes the National Forest System.

1906    Passage of the Antiquities Act gives the president the authority to declare national monuments on public lands in order to protect sites of historic or scientific interest without congressional approval.

| | |
|---|---|
| **1908** | U.S. Supreme Court establishes the doctrine of a federal reserves water right in *Winters v. United States.* |
| **1910** | Bureau of Mines is created as a research-oriented agency to collect information on supply and demand, mine safety, and environmental issues. |
| **1911** | U.S. Supreme Court rules that the federal government has the right to create a national forest on federal public land without consent of the state within which the forest is located in *Light v. United States.* |
| **1911** | Weeks Act gives specific legislative authority to the Division of Forestry for forest protection, enhancing the agency's jurisdiction and power. The statute also gives consent of Congress for states to enter into agreements to conserve forests and the water supply. |
| **1912** | Federal officials debate the use of automobiles in national parks at a conference in Yosemite National Park, and consider issues of roads, accidents, conflicts with horse-drawn carriages, and limitations on where automobiles should be allowed to travel. |
| **1915** | The Santa Fe and Union Pacific railroads spend $500,000 to set up exhibits in national parks, encouraging Americans to ride the rails to the parks. The alliance is considered part of the efficient, economic, and satisfactory management of the Forest Service as advocated by Gifford Pinchot. |
| **1916** | National Park Service is established within Department of the Interior to protect scenic areas, historic objects, and wildlife for future generations, providing a management structure for the units that had previously been run independently. |
| **1916** | Congress amends Lacey Act to reduce penalties for hunting, destroying timber, or removal of minerals in Yellowstone National Park. |

1917    Congress gives secretary of agriculture the authority to use a leasing system for hardrock minerals on acquired lands.

1918    First National Park Service regulations are published, and include maintaining the parks "in absolutely unimpaired form for the use of future generations as well as those of our own time." Regulations also outlaw leasing for summer homes, cutting of trees except for buildings and where it would not hurt the forests, and requiring roads to harmonize with the landscape.

1919    Steven Mather, "father" of the national parks, complains of too much development and commercialization in the park system.

1920    Mineral Leasing Act provides that federal lands containing deposits of coal, phosphates, oil, and other minerals may be acquired only through a leasing system.

1920    Federal Power Act authorizes construction of dams on federal lands.

1921    Congress hastily enacts amendments to the Federal Power Act to forbid the building of dams in national parks and monuments without its specific approval.

1921    Aldo Leopold envisions wilderness as a "continuous stretch of country preserved in its natural state, open to lawful hunting and fishing, devoid of roads or other works of man."

1922    National Park Service conference in Yosemite produces a resolution on development, stating that it is undesirable and should be avoided, and that plans for the development of each national park should be outlined as far in advance as possible.

1922    At the urging of Aldo Leopold, 574,000 acres in the Gila National Forest in New Mexico are set aside for wilderness recreation.

1924    National Conference on Outdoor Recreation meets to discuss the need for public use of forests and other public lands for recreation, leading to expansion of national park system.

1924    Clarke-McNary Act authorizes the secretary of the Interior to recommend systems of fire prevention and suppression for each forest region, and increases federal fire-prevention funding authorization.

1926    U.S. Forest Service inventories public lands that are roadless and manages them as wilderness areas.

1927    Financier John D. Rockefeller, Jr., offers $5 million in matching funds to create Great Smoky Mountains National Park. Sufficient matching funds are not raised until 1934.

1928    Congress defines how concessions are to be managed within the national parks.

1929    Forest Service begins to establish "primitive areas," wilderness-type areas managed under flexible rules with little scientific study, and assumed to be withdrawn from development only temporarily.

1930    Bob Marshall publishes his influential essay "The Problem of the Wilderness" in *Scientific American* magazine, beginning a decade-long crusade for wilderness preservation.

1931    National Park Service approves a forestry policy which states that the parks will be as completely protected as possible, including a fire protection plan, control of insects, tree disease control, campground protection, land and timber exchanges, cleanup of dead and downed timber, and the elimination of grazing as soon as practicable.

1932    Wildlife conservationist George Wright releases a report that lays the ground rules for scientific wildlife management and further research.

1933        Civil War battlefields are added to the National Park system, expanding the network of sites from scenic and wild areas to historic resources.

1933        Creation of the Civilian Conservation Corps (CCC) by Franklin D. Roosevelt leads to construction of an extensive infrastructure system within the national parks, including roads, trails, and other public works projects. CCC workers take on responsibility for fire control and timber management.

1934        Congress enacts Taylor Grazing Act in response to problems caused by severe overgrazing and damage to soils, and creates Grazing Division.

1935        President Franklin Roosevelt establishes the National Resources Board by executive order, to prepare and present to the president a program for the development and use of land, waters, and other national resources.

1935        The Wilderness Society is founded to protect vast tracts of public lands from natural resource extraction and development. Its founders include the nation's foremost advocates for preservation Aldo Leopold, Bob Marshall, Benton MacKaye, Harvey Broome, Bernard Frank, Harold Anderson, Ernest Oberholtzer, and Robert Sterling Yard.

1935        Congress declares that it is national policy to preserve for public use historic sites, buildings, and objects of national significance, greatly expanding the jurisdiction of the National Park Service.

1935        National Park Trust is established to provide a mechanism for the private sector to donate land, gifts, and funds to the parks.

1936        Congress enacts the Park, Parkway, and Recreational Area Study Act to aid state and local agencies in recreational planning.

| | |
|---|---|
| 1939 | Grazing Office is renamed the Grazing Service as part of an expansion of federal regulation of public lands grazing. |
| 1942 | Cooperative Forest Fire Prevention Campaign begins as part of the patriotism fueled by the war effort, assisted by the Wartime Advertising Council and professional advertising agencies. |
| 1942 | Congress terminates the Civilian Conservation Corps as national park attendance drops, and the nation's attention turns to World War II. |
| 1945 | Smokey Bear first appears as a symbol of the nation's fire prevention effort. |
| 1946 | Bureau of Land Management is created through the merger of the General Land Office and the Grazing Service, with responsibility for managing all remaining "unreserved" public lands. |
| 1950 | National Park Service officials oppose the building of Echo Park Dam in Dinosaur National Monument, a position contrary to that of the Department of the Interior. |
| 1950 | Grand Teton National Park is established in Wyoming, provided that no further extension or establishment of national parks or monuments occur within the state without congressional approval. |
| 1952 | Chief of U.S. Forest Service tells Congress that without funds being appropriated to build new roads, many areas of valuable timber will remain undeveloped. |
| 1953 | *Harper's Magazine* journalist Bernard DeVoto writes a scathing article criticizing Congress for ignoring the national park system and treating the National Park Service like an "impoverished stepchild." DeVoto cites the deplorable conditions of the park system's |

roads, campgrounds, and buildings, and complains the parks are understaffed.

**1954**   Multiple Mineral Development Act enacted to deal with hardrock mineral claims that are being leased for oil.

**1954**   James B. Gilligan delivers a speech before the Society of American Foresters warning that the country has become complacent about wilderness preservation, and that congressional action is necessary. The speech reignites the wilderness preservation advocacy groups, which had been almost dormant during World War II.

**1955**   Surface Resources Act restricts the surface use of mineral claims to uses required for mining purposes.

**1956**   National Park Service director Conrad Wirth presents Mission 66 to President Dwight D. Eisenhower, a ten-year program to prepare a plan for an estimated 80 million visitors by 1966, with the paramount goal of preserving wilderness areas in the parks.

**1956**   Proposal to dam the Green River in Dinosaur National Monument prompts Howard Zahniser of the Wilderness Society to draft a wilderness protection law.

**1957**   Senator Hubert Humphrey formally introduces a wilderness bill, which will not receive sufficient support for passage until 1964.

**1960**   Multiple Use and Sustained Yield Act widens access and use of public lands by requiring that they be managed for a variety of uses that best meet the needs of the American people, a theme that guides future development of natural resources in the United States.

**1962**   President John F. Kennedy, addressing the White House Conference on Conservation, emphasizes the importance of the role of government in maintaining environmental quality and natural beauty.

1963        Interior Secretary Stewart Udall proposes a "New Conservation" that would make national parks "spacious areas of superior scenery to be preserved forever," with a doubling of the acreage under National Park Service control by 1970.

1963        Leopold Report, initially designed to comment on wildlife issues in the national parks, includes recommendations on a basic management philosophy and calls for a permanent staff of scientists to oversee the maintenance or restoration of parks to the conditions that existed prior to European settlement.

1963        Recreation Advisory Council recommends that greater efforts be made to fulfill the increasing demand for outdoor-recreation areas.

1964        Congress enacts the Wilderness Act and calls for designation of undeveloped federal lands to be kept in their natural state for the use and enjoyment of the public. More than 9 million acres are immediately designated as wilderness areas.

1964        Classification and Multiple Use Act gives the Bureau of Land Management temporary authority to manage public rangelands.

1964        U.S. Bureau of the Budget publishes a study, *Natural Resource User Charges*, which shows that the federal government should receive a fair price from public land resource use, including livestock grazing.

1964        Public Land Law Review Commission is established to provide a comprehensive review and evaluation of public land policies.

1964        Lyndon B. Johnson of Texas wins presidential campaign as the only candidate to recognize the political value of the environmental movement, using the concepts of the "New Conservation" and "natural beauty" to appeal to middle-class Americans.

1965    Land and Water Conservation Fund Act is enacted to provide the federal government with funds to purchase additional public lands. Funds are generated through receipts from oil and gas leases.

1965    Concessions Policy Act establishes long-term, renewable monopoly contracts for building and commercialism in national parks, addressing the controversial issue of private enterprise in the parks.

1966    National Historic Preservation Act authorizes the secretary of the Interior to create and maintain a national register of historic districts, sites, and structures, and to establish programs of matching grants, with the National Park Service becoming the coordinating agency for the designated areas.

1966    Department of Transportation Act declares that it is national policy to preserve the natural beauty of the countryside, park and recreation lands, wildlife and waterfowl refuges, and historic sites. The law expands responsibility for resource conservation to a non–land management federal agency.

1967    National Park Foundation is established by Congress as the successor to the National Park Trust Fund, with assets of $784,387.

1968    U.S. departments of Interior and Agriculture propose a grazing fee schedule that would require the cost of grazing on public lands to equal that of private lands, to be phased in over a ten-year period.

1968    National Wild and Scenic Rivers Act and the National Trails System Act greatly expand and diversify the national park system and create complicated management problems.

1968    Atlantic Richfield Company discovers superfield of oil on the North Slope in Alaska.

1968    National Park Service develops a policy for the use of fire as part of the ecosystem management plan for Sequoia and Kings Canyon national parks, establishing prescribed fire as a tool for restoring areas to their natural condition, and the acceptance of active management of the natural environment.

1970    First observance of Earth Day becomes the peak event for the environmental movement, as an estimated 20 million citizens participate in marches and protests, educational events, and public outreach.

1970    Public Land Law Review Commission issues recommendations on existing public land laws in *One Third of the Nation's Land.*

1970    Passage of National Environmental Policy Act requires that all federal projects be reviewed to assess the potential environmental impact, including alternatives to the proposed federal action, with a requirement for public participation. It also establishes a Council on Environmental Quality to assist and advise the president on environmental policies. The legislation is neither extensively lobbied nor thoroughly debated as timber, mining, and other industry interest groups understand the implications of the law.

1970    Grazing fee increase moratorium goes into effect, prohibiting any increase over the fees charged in 1969.

1970    Passage of the Clean Air Act provides incentives for increased production of Western coal, which is lower in sulfur and does not produce as much air pollution as Illinois basin coal.

1970    Congress enacts the General Authorities Act and specifies that all units be administered by the National Park Service as part of the same system, to be managed according to the provisions of the Organic Act of 1916.

1970     A riot in Yosemite's Stoneman Meadows by counter-
         culture youth spurs the National Park Service to begin
         a comprehensive law enforcement training program,
         broadening the responsibilities of park rangers.

1970     Bolle Report calls for changes in management of U.S.
         forest management practices and criticizes clear-cut-
         ting.

1971     President Richard Nixon uses the presidential power
         of executive order to strengthen the process of historic
         preservation by expanding and maintaining the Na-
         tional Register of Historic Places. The provisions of
         the executive order are subsequently included in the
         1980 National Historic Preservation Act amendments.

1971     Environmental groups challenge the U.S. Forest Ser-
         vice's liquidation of old growth forests in Alaska's
         Tongass National Forest in *Sierra Club v. Hardin.*

1971     Forest Service establishes policy to allow lightning-
         caused fires in wilderness areas to burn under specific
         conditions, adopting term "prescribed fires of natural
         origin."

1971     Council on Environmental Quality intervenes in the
         controversy over a proindustry Timber Supply Act
         and attempts by the Forest Service to accelerate log-
         ging and road building in potential wilderness areas,
         thereby disqualifying them from wilderness designa-
         tion.

1972     Study of Northern spotted owl and Pacific Northwest
         forests conducted by Oregon student initiates conflicts
         over harvests of old growth forests.

1972     U.S. Forest Service conducts Roadless Area Review
         and Evaluation (RARE) to survey areas for possible
         inclusion in the National Wilderness Preservation
         System.

| | |
|---|---|
| 1972 | Two specifically urban recreation areas are added to the profile of the national parks with the addition of the Gateway National Recreation Area in New York and the Golden Gate National Recreation Area in San Francisco. |
| 1972 | An activist group, the Conservation Foundation, issues a blueprint of how national parks should be managed to best preserve the resources they were designed to protect, addressing issues such as park concessions, visitor capacity, and research. |
| 1973 | Arab nations angry at U.S. support of Israel in 1973 war institute oil embargo against the United States and Holland. Organization for Petroleum Exporting Countries (OPEC) decreases production, and six months later, world crude oil prices triple to $12 per barrel. OPEC is firmly established as being in control of world oil market. |
| 1973 | OPEC imposes a sixfold increase in prices for crude oil, triggering widespread shortages and price gouging. President Nixon announces a sixty-day freeze on gasoline prices. |
| 1973 | Izaak Walton League, U.S. environmental organization, sues U.S. Forest Service over clear-cutting. |
| 1973 | Eastern Wilderness Act allows areas that have recovered, or are in the process of recovering, from past use to be included in the National Wilderness Preservation System, providing coverage for areas in the East that are no longer pristine. |
| 1974 | Federal court decision in *Natural Resources Defense Council v. Morton* requires Bureau of Land Management to conduct site-specific environmental impact statements instead of one covering its entire grazing program. |
| 1974 | Forest and Rangeland Renewable Resources Planning Act requires agencies to assess conditions in all forests |

and grazing areas, and to develop a strategic plan for their management.

1974    President Richard Nixon resigns after Watergate scandal; President Gerald Ford pushes Project Independence to make the United States independent of petroleum imports by 1985.

1974    Final RARE report recommends that 12.3 million acres of U.S. Forest Service lands be designated for wilderness protection.

1975    A second moratorium on grazing fee increases is imposed by the U.S. departments of agriculture and interior.

1975    Energy Policy and Conservation Act establishes price incentives, establishes Strategic Petroleum Reserve, and creates Corporate Average Fuel Economy Standards (CAFE) in an attempt to insulate the United States against world oil-production fluctuations.

1975    Congress enacts the Eastern Wilderness Act, designating sixteen additional wilderness areas, expanding the area of advocacy concerns from the West to the East.

1976    Federal Land Policy and Management Act gives Bureau of Land Management mandate to implement multiple use principles, and extends Wilderness Act provisions to public domain lands managed by the agency. Grazing fees are frozen at 1969 levels.

1976    Federal Coal Leasing Amendments Act provides for systematic study of energy sources and calls for competitive leasing of lands for mining and energy prospecting.

1976    National Forest Management Act calls for development of a strategic plan for each of the national forests, which eventually limits controversial logging practices such as clear-cutting.

1977    Congress enacts the Surface Mining Control and Reclamation Act, giving additional funding to producers, and establishes the Office of Surface Mining and Reclamation to deal with surface mining of coal.

1977    Department of Energy is created under President Jimmy Carter to develop a comprehensive energy plan by consolidating many of the functions and responsibilities of other federal agencies. Department of the Interior loses most of its jurisdiction relating to energy policy.

1978    Interior secretary Cecil Andrus temporarily withdraws 110 million acres of Alaska public lands from development in response to pressure from environmental groups.

1978    McClure Amendment to the Public Rangelands Improvement Act limits grazing reductions in the West to 10 percent per grazing permit.

1978    Uranium Mill Tailings Reclamation Act provides funds to identify and move or bury uranium waste near residences.

1978    Congress places moratorium on grazing fee increases for the fourth time since 1969.

1978    Revolution in Iran results in a major drop in crude oil production, and the United States considers options for drilling on the North Slope area of Alaska.

1978    Congress expands the boundaries of California's Redwood National Park to protect stands of coastal redwoods, the watershed, and the ecosystem, agreeing to compensate those unemployed by the loss of timber-related jobs.

1978    Cooperative Forestry Assistance Act rewrites provisions of the 1924 Clarke-McNary Act, emphasizing rural fire prevention and control rather than forest fires.

1980    Alaska National Interest Lands Conservation Act transfers federally held land to the state of Alaska and to Native American tribes, and moves responsibility for land management from the Bureau of Land Management to the Fish and Wildlife Service and National Park Service. The statute more than doubles the land managed by the National Park Service.

1980    President Ronald Reagan appoints James Watt as secretary of Interior in an attempt to reorient range management decision-making to favor resource production goals over environmental protection.

1980    New Bureau of Land Management director Robert Burford promises to adopt a "good neighbor" policy that places greater emphasis on the economic health of extractive industries.

1980    Congress enacts National Materials and Minerals Policy, Research and Development Act, advocating a more coherent and coordinated minerals policy.

1980    National Park Service's *State of the Parks* report identifies specific threats that endanger park resources, and calls for a comprehensive inventory of the natural and cultural resources of each park, monitoring programs, and an expansion of the research and resource management staff, which the reports says are "inadequate to respond to the needs" of the agency.

1980    President Jimmy Carter supports Synthetic Fuels Act to offer substantial incentives and subsidies to industry to encourage the development of domestic fuels.

1980    General Management Plan for Yosemite National Park calls for removal of all automobiles and redirection of development to the periphery of the park and beyond.

1982    Minerals Management Service is created to administer leases on the outer continental shelf and manage royalty and revenue functions on the public lands.

1982    Concerns about natural resource management make their way to the ballot box, as two-thirds of the candidates endorsed by the League of Conservation Voters are elected. The Reagan administration admits that it may have misread the previous election's mandate on environmental issues.

1982    Nuclear Waste Policy Act is enacted to find two suitable sites, one in the East and one in the West, as permanent repositories for high-level radioactive waste.

1984    National Security Minerals Act passes, to coordinate mineral policy related to national defense.

1985    Food Security Act establishes the Conservation Reserve Program, to expand federal subsidies for forest reforestation, including funding for private owners of farms and ranches.

1985    Iran-Iraq War causes a major drop in crude oil production and an increase in price from $14 a gallon to $35.

1987    Sierra Club Legal Defense Fund opens new offices in Seattle to challenge governmental policies relating to old growth forests.

1987    Congress amends Nuclear Waste Policy Act to restrict the search for a nuclear waste storage facility to a single site.

1987    Federal government's General Accounting Office issues a report which states that little progress has been made in documenting and mitigating threats to the national parks, including threats to parks' aesthetic qualities, cultural resources, air and water quality, plants, and wildlife.

1988    Former Secretary of Interior Stewart Udall and others found the Mineral Policy Center, with a goal of reforming the 1872 General Mining Law.

1988    Massive fires burn 1 million acres in Yellowstone National Park and 600,000 acres in adjacent national forests, leading to controversy over park management and wildfire practices. National Park Foundation establishes Yellowstone Recovery Fund to raise funds to help the park recovery effort.

1988    Wise use conference in Reno, Nevada, brings together mineral prospectors, ranchers and livestock owners, motorized vehicle owners, and property rights advocates to coalesce around land use issues.

1988    Oregon and Washington list Northern spotted owl as a threatened species, putting pressure on the federal government to follow suit.

1989    Association of Forest Service Employees for Environmental Ethics is established to challenge agency actions such as timber harvests.

1989    U.S. District Court judge William Dwyer rules that U.S. Forest Service plans to protect the Northern spotted owl, designated as a threatened species, are inadequate, issuing injunctions against timber sales in Oregon and Washington.

1989    Contentious hearings in Congress debate the natural burn policy used in Yellowstone National Park, condemning the economic impact during the tourist season.

1990    Congress enacts Tongass Timber Reform Act as the environmental lobby battles timber industry proposals.

1990    General Accounting Office report on changes in federal fire management made after the Yellowstone fires is taking longer than expected, citing interagency cooperation as a problem.

1991    Polls taken by the Roper Organization indicate that the National Park Service has the highest approval rating of all government agencies.

1991    Saddam Hussein invades Kuwait to gain control of the country's oil resources; the United States becomes involved to maintain strategic access to Middle East oil reserves. Iraqi forces set fire to nearly 800 Kuwaiti oil wells, irrevocably damaging the country's ecological landscape.

1991    Catron County, New Mexico, passes an ordinance requiring all federal and state agencies to comply with its own land-use policy plan in an effort to diminish federal control over public lands.

1991    Federal judge halts all timber sales until the federal government complies with forest legislation.

1991    Dissolution of the Union of Soviet Socialist Republics opens up the region to scrutiny, revealing unimaginable environmental degradation, including contaminated soil and water, a lack of security near nuclear power facilities, and uncontrolled well fires in the Caspian region. The changes also lead to heightened interest by Western energy companies that previously had been shut out of the area.

1992    President Bill Clinton appoints Bruce Babbitt as secretary of Interior. Babbitt identifies public land reform as his top policy priority.

1992    Chief Forester Dale Robertson announces that ecosystem management is the new concept guiding U.S. Forest Service decision-making.

1992    U.S. District Court judge William Dwyer rules that in the U.S. Forest Service *Thomas Report,* the agency did not address issues relating to species other than the spotted owl.

1992    Passage of the Energy Policy Act imposes a drilling ban in the Arctic National Wildlife Refuge in exchange for easing restrictions on the licensing of nuclear power plants.

**1992**     Vail Agenda outlines the state of conditions in the National Park Service after seventy-five years and makes recommendations for the reform and rejuvenation of the agency, including a need for additional resources, an extensive training program, more use of exhaustive scientific research, and the use of park unit surveys in the management of programs.

**1992**     Office of Technology Assessment report finds that multiple-use and sustained-yield issues are inherently conflicting, incompatible, or exclusive.

**1993**     Shell Oil shuts down its operations in Nigeria as a result of protests over pollution, poor clean-up efforts, and increasing violence. A class action suit is filed in New York against ChevronTexaco under the Alien Tort Claims Act on behalf of an estimated 30,000 Amazon residents for alleged pollution of their environment.

**1993**     Nye County, Nevada, passes an ordinance to overturn federal authority over public lands within the county by attempting to reopen a road closed by the U.S. Forest Service in the Toiyabe National Forest.

**1993**     President Bill Clinton convenes the Northwest Forest Summit in Portland, Oregon, to bring together stakeholders in the debate over Northwest forest policies and the fate of the Northern spotted owl, a threatened species.

**1993**     Under an executive order, the Clinton administration establishes the National Biological Survey (later renamed the National Biological Service), to foster an ecosystem management approach to address land management issues.

**1993**     British Petroleum locates oil beneath the eastern plains in the Andes Mountains of Colombia, predicting millions of dollars in exports for the country.

| | |
|---|---|
| **1994** | Secretary of the Interior Babbitt abandons legislative strategy in favor of administrative regulations with Rangeland Reform '94. Babbitt later announces that he is ending his effort to change the formula by which grazing fees are calculated. |
| **1994** | Republicans gain control of Congress in midterm elections and select supporters of existing rangeland policies to chair major natural resource committees. |
| **1994** | U.S. Forest Service cancels two timber sale contracts in Alaska's Tongass National Forest, ending long-term agreements originally designed in the 1950s to stimulate economic development. |
| **1994** | Government officials in Azerbaijan agree to allow consortium of foreign oil companies to invest $8 billion for oil production over a thirty-year period. |
| **1995** | Western Republican members of Congress unsuccessfully attempt to pass legislation to establish livestock grazing as the dominant use of federal lands. |
| **1995** | Rescissions Act (also known as the Timber Salvage Rider) is developed under the Clinton administration, opening up areas of old growth forest that had previously been protected by exempting timber salvage sales from legal challenges, insulating many timber sales from citizen appeals and environmental reviews. |
| **1995** | Congress abolishes the Bureau of Mines. |
| **1995** | National Coal Association and American Mining Congress merge to create the National Mining Association, to oppose changes in mining laws. |
| **1995** | Senate passes a bill to designate livestock grazing as a dominant use of the public lands, but the legislation fails to get sufficient support in the House of Representatives. |

1995     Wolves are reintroduced to Yellowstone National Park as part of the restoration of an "endangered species."

1996     President Bill Clinton uses the 1906 Antiquities Act to establish the 1.7-million-acre Grand Staircase-Escalante National Monument by executive order, bypassing Congress and public participation.

1996     Congress authorizes a three-year Recreational Fee Demonstration Program to test new or increased entrance and user fees in units managed by the National Park Service, Bureau of Land Management, Forest Service, and Fish and Wildlife Service. At least 80 percent of fees are to be retained at the site where they are collected, supplementing agency funds rather than offsetting appropriations.

1997     Antilogging activist Julia "Butterfly" Hill begins a two-year "tree sit" in northern California to protest the harvesting of old growth trees.

1997     Forest Service appoints a committee of scientists to review land and resource management planning processes.

1997     Wyoming Farm Bureau files suit in federal district court to halt wolf reintroduction in Yellowstone National Park and loses in the appeal process.

1997     U.S. General Accounting Office tells Congress that the U.S. Forest Service's decision-making process is "broken and in need of repair."

1998     China bans logging but timber imports skyrocket, much of it from illegal logging in Indonesia.

1998     California gubernatorial candidate Gray Davis pledges that if he is elected, "all old growth trees will be spared from the lumberman's axe." Davis wins the election, but as governor, does not keep his electoral promise. He is subsequently recalled from office and loses a special election to Arnold Schwarzenegger.

| | |
|---|---|
| **1998** | Western Governors' Association initiates Enlibra model, to give those living near natural resources more influence on land management decisions. |
| **1999** | UN Security Council acts to enforce sanctions in diamond sales by rebel groups during the civil war in Angola. |
| **1999** | Report by the National Research Council concludes that better enforcement and coordination of existing mining regulations are better than new regulations. |
| **1999** | Clinton administration proposes an eighteen-month ban on new road building in roadless areas to protect national forests and ecosystems. |
| **1999** | Yana Curi Report details the impact of oil development on the health of the people in the Ecuadorian Amazon who live close to oil fields. |
| **2000** | Public Lands Council, a trade organization representing ranchers, loses a U.S. Supreme Court case challenging Department of Interior rangeland regulations. |
| **2000** | Organization for Petroleum Exporting Countries tightens exports, leading to soaring prices for oil and gasoline. In 1999, a barrel of oil sells for $10; the price triples in 2000. |
| **2001** | Ecuadorian government grants approval for the start of construction of a new $1.3 billion, 298-mile oil pipeline through the rain forest that is expected to double the country's oil production. |
| **2001** | Tengizchevroil, the ChevronTexaco-led international consortium developing an oil field in western Kazakhstan, is fined $75 million for causing ecological damage. |
| **2002** | President George W. Bush announces his Healthy Forests Initiative, to protect communities from wild- |

fires, but environmental groups contend that it allows for increased logging.

2003    Twenty-six Tagaeri tribesmen in Ecuador are massacred, allegedly by Colombian logging interests involved in the illegal timber trade.

2003    President Bush signs the Healthy Forests Restoration Act to broaden Forest Service authority to remove hazardous fuels and open timber sales in wildfire-prone areas.

2004    U.S. Forest Service releases a new rule allowing forest managers to use an Environmental Management System approach for deciding the best use of forests on a case-by-case basis.

2004    Bolivian voters support a referendum endorsing the development of the country's gas reserves; the indigenous population doubts that they will see the benefits of development.

2004    United Nations reports that by 2025, one out of two Africans will be living in countries facing water scarcity.

2004    Federal Lands Recreation Enhancement Act extends Recreational Fee Demonstration Program and authorizes implementation for ten additional years.

2004    A forest activist is arrested three hours after climbing up a China fir tree on the grounds of the state capitol in Sacramento, California, as part of a protest against the logging of old growth forests.

2004    Ecuador approves plans by Brazil's state oil company to drill in Yasuni National Park, designated as a UN-protected biosphere reserve in 1989.

2005    Hurricanes Katrina, Rita, and Wilma devastate the Gulf Coast and Florida, leading to soaring gas and home-heating-oil prices throughout the United States.

**2005**     Court decision in *Earth Island Institute v. Ruthenbeck* requires Forest Service to allow for public notice, comment, and appeal of all categorical exclusions in national forests, voiding previous regulations.

**2006**     Oil prices top $70 per barrel in response to uncertainty in the Middle East, political uprisings in Venezuela and Nigeria, and concerns about growing consumption in China.

# 5

# Biographical Sketches

## Ansel Adams (1902–1984)

The policy-making process is affected by a number of factors, from political leadership and public opinion to news media coverage and protests. What is not often recognized, however, is the impact of artists, regardless of the tools they use. Ansel Adams's photography, often spare black and white images, has had a profound effect on Americans' attitudes about wilderness and protected areas that in some ways reaches a broader audience than most organizations.

Adams's father was an entrepreneur who started several businesses but who also homeschooled his only son, who was born in 1902, just before the San Francisco earthquake. At age fourteen Adams and his parents took a vacation in Yosemite National Park, a trip that would from then on become a yearly one. The experience was life changing, giving Adams an appreciation of the Sierra Nevada that would transform into advocacy later in his life. In 1919 he got a job as a custodian at the Sierra Club's Le Conte Memorial Lodge in Yosemite, and nine years later he married the daughter of the owner of a store and studio there.

The work at Yosemite was not included in Adams's original career plans. He had studied to become a professional pianist, a dream he did not abandon until age twenty-eight. He also became an apprentice to the owner of a photofinishing plant in San Francisco, learning the intricacies of photography and gaining the use of a professional darkroom. He visited Yosemite, which he considered a holy place, bringing along his bulky camera equipment and developing his own images of the park. In 1927 he became the Sierra Club's official trip photographer, and his work was published regularly in the organization's publications. He was elected to the group's board of directors in 1934 and remained on the board for thirty-seven years.

His primary contribution was the memorable imagery of his photographs, which showed the country a beauty that had never been captured on film or by other artists. He lobbied presidents in person and wrote passionately about the need to protect natural areas. For his devotion to preserving the country's pristine wilderness, he was awarded the Conservation Service Award by the Department of the Interior in 1968, and the Presidential Medal of Freedom in 1980. When he died of a heart attack at age eighty-two, his ashes were scattered on the slope of Mount Ansel Adams, located on the boundary of Yosemite, his spiritual home.

# Bruce Babbitt (b. 1938)

In Arizona, the Babbitt name is as close to a founding family as one could get. The Babbitts settled northern Arizona in the 1880s, building a ranching and mercantile business and operating trading posts on the Navajo Reservation. Bruce Babbitt was one of six children, leaving home to attend college at the University of Notre Dame, graduating in 1960 with a major in geology. He continued his education in England, where he received a master's degree in geophysics in 1963. He then went to Harvard Law School, and after graduation worked with civil rights and antipoverty programs, returning to Arizona in 1968 and joining a law firm.

Babbitt's political career began in 1975, when he became Arizona attorney general, and then, in 1978, became governor when his predecessor resigned. The position allowed Babbitt to create the state's Department of Environmental Quality to spearhead land reform, and to write a comprehensive groundwater code. His commitment to social justice and environmental protection drew praise from national Democratic Party leaders, and a brief candidacy for president in 1988. For the next five years he served as the director of the League of Conservation Voters and worked toward passage of a state ballot initiative that earmarked lottery funds to acquire and maintain parks and wilderness areas in the state.

His environmental experience garnered the attention of newly elected president Bill Clinton, who appointed Babbitt secretary of the Department of the Interior in 1993, a position that he held until Clinton left office in January 2001. Babbitt's tenure within the agency was controversial from the beginning. Environmental group leaders who expected him to be an advocate for the preservation of public lands were disappointed at initiatives they considered too oriented toward the concept of multiple use. Supporters point to his ability to bring together coalitions that resulted in the signing of the California Desert Protection Act, restoration of the Florida Everglades, and his efforts to establish more than 6 million acres of land for national monuments. He received positive feedback for programs to restore wolves and the California condor to their natural habitat, although the reintroduction angered ranchers. He was less successful in lobbying for the creation of the National Biological Survey, designed to bring additional scientists into land management, and his Rangeland Reform '94 program, which would have significantly raised grazing fees.

When Clinton left office, Babbitt continued to work as a consultant and writer, promoting his book *Cities in the Wilderness*. He continues to live in Washington, D.C., although he frequently returns to northern Arizona, still home to many members of the Babbitt family after more than a century.

# David Ross Brower (1912–2000)

When David Brower died at age eighty-eight, he left behind a legacy of environmental crusades that inspired, encouraged, and angered activists throughout the United States. But he was also a central figure in the evolution of the environmental movement, shaping policies that would last long after his death.

From his home in Berkeley, California, Brower was introduced to hiking and nature as a child. He started as a student at the University of California but dropped out in 1929 without finishing a degree. His antipathy toward academia was lifelong, and he frequently railed against scholars who spent too much time in their offices instead of in the field. Brower was first known as a mountaineer who developed climbing routes across many of the nation's top sites in the 1930s. His interest in mountaineering drew him to the Sierra Club, which he joined in 1933. At the time, it was a hiking club without a political agenda. He wrote articles for the group's newsletter, campaigned for the inclusion of Kings Canyon as a national park, and in 1941 became a member of the Sierra Club's board of directors. He became its first executive director in 1952, and saw its membership grow from 2,000 to 77,000 by the time he left the organization in 1969. He helped form the Sierra Club Foundation in 1960, and successfully managed the fight to block the building of a proposed dam in Dinosaur National Monument. He once said that his life's major disappointment was a compromise agreement he made with the federal government to allow a dam to be built in Glen Canyon in exchange for the demise of the dam proposed at Dinosaur.

Working in conjunction with the Wilderness Society, Brower expanded the Sierra Club's efforts to preserve wild places, lobbying in support of the 1964 Wilderness Act. The speed at which he sought change proved too rapid for many of the members of the Sierra Club's board of directors, who forced him to leave the organization. He founded a new group, Friends of the Earth, in 1969, expanding the environmental agenda even more by including international issues. He also helped found the League of Conservation Voters, another vehicle for mobilizing the public by analyzing the environmental voting records of members of Congress. Once again his vision was not shared by the group's leadership, and he was fired. Moving on, he served as chair of the board of directors of another new group he founded, Earth Island Institute.

Brower left behind dozens of markers of his influence, and he was twice nominated for the Nobel Peace Prize. The biennial Fate and Hope of the Earth Conferences, the Global Conservation, Preservation, and Restoration (CPR) Service, plans for the creation of a National Biosphere Reserve System, and numerous parks, seashores, recreation areas, and wilderness lands are a result of his tireless efforts.

# Robert Marion Clawson (1905–1998)

Marion Clawson, one of the nation's preeminent forest policy experts, grew up in a part of the West where trees and timber were considered less important than ranching and mining. He was born in Elko, Nevada, and attended high school there, later receiving his bachelor's and master's degrees from the University of Nevada, Reno. His interests in agriculture led him to study the economics of Nevada's farms while in college, and from 1929 to 1946 he worked for the Bureau of Agricultural Economics within the U.S. Department of Agriculture (USDA). He coordinated research for the USDA and for the Bureau of Reclamation within the Department of the Interior, conducting an investigation of the Columbia Basin in Washington and the Central Valley Project in California. In 1943 he was awarded a Ph.D. in economics from Harvard.

Clawson served as the second administrator of the Bureau of Land Management (BLM), from 1948 to 1953. He often bragged that he was fired from his job at the BLM when a new Republican administration was elected. He later served as a consultant and economic adviser in several other countries, including Israel, Venezuela, Chile, Pakistan, and India. In the United States, he joined Resources for the Future (RFF) in Washington, D.C., and served in a consultant capacity for the United Nations and the Rockefeller and Ford foundations. His academic career included short-term appointments at the University of California, Berkeley, the University of Washington, and at Duke University. He was a member, director, or founder of a wide range of organizations, including the American Academy of Arts and Sciences, the Forest History Society, the Range Management Society, and the Soil Conservation Society.

His professional affiliations and prolific career as a writer did not begin to focus on forestry and forest policy until the early 1970s, when he served as a member of the President's Advisory Panel on Timber and the Environment. Clawson became both a critic of the U.S. Forest Service (FS) and a proponent of economic efficiency in forest management—what might be called a "wise use conservationist"—along the model of the chief of the FS, Gifford Pinchot. His first books, from the late 1940s and early 1950s, were about agriculture; his last book, written in 1987, was his memoirs, and he continued to work at RFF on a regular basis until his death in 1998 at age ninety-two. As one colleague notes, "Marion never really retired." Clawson ended his career as both a researcher and public servant, having testified before Congress, taught university students, advised foreign governments, and reorganized one of the government's major environmental agencies—major accomplishments for the son of a Nevada rancher and miner.

# Bernard DeVoto (1897–1955)

Although DeVoto is considerably less well known than many of the other environmental leaders of his time, Benny DeVoto deserves a special place in U.S. history as one of the preeminent writers of his time who focused attention on the federal government's management (or mismanagement, as he termed it) of public lands.

DeVoto was born in Ogden, Utah, just after the region had officially been given statehood. His parents were of mixed religious persuasion: his father was from a family of Catholics, while his mother's family were Mormons. He spent only one year at the University of Utah, quitting in disgust when two members of the faculty were fired and the school was censured. He applied to Harvard, becoming what one biographer has called "one of the most Eastern Westerners who ever was." His stay at Harvard was interrupted when he enlisted in the infantry, but he returned to finish his degree in philosophy in 1920 and gained a reputation there as a talented and confident writer. DeVoto returned to Ogden, suffering from severe depression and insomnia for a year before accepting a position teaching English at Northwestern University. He married in 1923 and moved to Cambridge, Massachusetts, to write novels and essays that appeared in many prestigious national publications, including the *Saturday Review of Literature,* where he was asked to serve as editor.

DeVoto moved on to *Harper's* magazine and took responsibility for the oldest running column in U.S. journalism, "The Easy Chair," and became a champion of consumers and eventually a crusader for numerous causes. After his 1946 tour of the West, he became an expert on land use and resource management, and between 1947 and the time of his death wrote more than forty articles about the region. He devoted articles and columns to condemning the excesses of overgrazing on public lands and the political power of the livestock industry. He told his readers how self-interested leaders were attempting to have public lands transferred to thirteen Western states with the right to dispose of them however they wanted. He called this the greatest land grab in history, with a few groups "hellbent on destroying the West."

A journalist used the Freedom of Information Act to discover that the Federal Bureau of Investigation had put together a 197-page file on DeVoto, much of it devoted to the writer's conservation work and as an "environmental agitator." Conservation groups, many of which were just beginning their role as wilderness advocates, distributed reprints of his columns; Secretary of Agriculture Clinton Anderson credited the author with single-handedly stopping the land grab. The author was awarded the Pulitzer Prize and the Bancroft Prize in 1947 for his book *Across the Wide Missouri,* and the National Book Award in 1952. DeVoto died of a heart attack at age fifty-eight in New York City, far from the West that had been the subject of so much of his journalistic venom.

# Wayne Hage (b. 1936)

Hage's name is synonymous with the phrase "Range War," and he is considered to be one of the leaders in the battle over grazing and private property rights in the West. Wayne Hage has waged a war against the federal government for decades, taking his case to the federal courts in a challenge to federal grazing laws and regulations.

What makes Hage unique among many of the sagebrush rebels who have fought against government regulation of public lands is that Hage has earned a reputation as a writer and self-taught historian. His book *Storm over Rangelands,* first published in 1989, has gained a cult following among wise-use advocates. He meticulously documents what he calls "the incessant attack on the western range livestock industry" that he says began as an assault by the conservation movement in the 1890s.

Hage's family had been involved with ranching for generations before he was born in Nevada, and he says that he always planned to continue the tradition. He attended the University of Nevada, Reno, and received several degrees there, including a master's degree in biological science. In 1978 he and his wife Jean (now deceased) purchased the Pine Creek Ranch in Nye County, Nevada, to graze cattle. He claims that the ranch is part of the "leftover land" that is too dry for farming but was conveyed to private individuals through a series of federal, state, and local laws, along with Western ranching customs.

In 1991, Hage filed suit in the Federal Court of Claims in Washington, D.C., against the U.S. Forest Service (FS), alleging that the agency had suspended and canceled his grazing permits, depriving him of grazing and water rights. The original complaint has been amended three times, and more than 200 pleadings have been filed in the case. The Bureau of Land Management (BLM) was later added to the suit as a codefendant; the suit said that the regulation of Hage's grazing activities resulted in a taking of a 752,000-acre surface estate, which he alleges includes the FS and BLM grazing allotments. Stewards of the Range, a nonprofit group founded in 1992 to pursue the Hage case, called this a violation of the constitutional prohibition against a "taking" without compensation guaranteed under the Fifth Amendment.

In 2002, the court ruled on the first of two issues: Hage was the rightful owner of certain water rights and ditch rights-of-way, with some associated forage rights. But the court rejected his claims regarding ownership of grazing permits and the surface estate, legally held by the federal government. The trial continued in 2004 to determine whether the government's actions are a taking, and if so, what compensation he is owed. It is still in litigation; the outcome will determine future controversies over grazing rights in the West.

# Julia Hill (b. 1974)

The 60,000-acre Headwaters Forest has been the site of bitter disputes over the logging of old growth redwood forests for decades. The area is home to the last stand of ancient trees, and is believed to represent about 3 percent of the original old growth forest that once stretched from the Big Sur coast in California to southern Oregon.

On December 10, 1997, a twenty-three-year-old activist climbed up a 200-foot-tall, 14-foot-diameter, 1,000-year-old coast redwood as part of a nonviolent "tree-sit." Protesters had often used the strategy of occupying trees that were in the path of loggers. Julia Hill, who would take the "forest name" Butterfly, sat on a tarp-covered platform about 180 feet above the ground, with food and other provisions hauled up and down by a series of ropes. She stayed in touch with supporters using a cell phone, balancing within the branches of the tree, which activists had called "Luna," for 738 days, climbing down on December 18, 1999.

Julia Hill was born on February 18, 1974, in Mount Vernon, Missouri, the daughter of a traveling evangelist minister who later brought his family to Jonesboro, Arkansas. She began attending high school there and later worked as a bartender in Fayetteville, Arkansas. In 1996 she was severely injured in an automobile accident, undergoing a year of treatment and recovery. Afterward, she considered various options and decided to travel with friends in the summer of 1997, winding up among other young people in the redwood forests of Humboldt County, California. She was so moved by the stories she heard about the logging of the old growth trees that she briefly returned to Arkansas, bought camping gear, and went back to California in November of 1997. She contacted the radical environmental group Earth First! and made her way to a base camp where other protesters were staying, where she heard about a planned tree-sit.

The protest began after the Pacific Lumber Company increased the amount of timber it was cutting on the 200,000 acres of land owned by MAXXAM Corporation. Earth First! had made the Headwaters Forest the focus of many of its activities, especially after the California Department of Forestry had approved a clear-cutting operation on steep slopes owned by MAXXAM. During the time she was involved in her protest, she endured harsh weather, attempts by timber company employees to get her to climb down, and criticism from those who believed that she was becoming a distraction to the environmental movement's broader goals. She was interviewed by the news media (some of whom climbed up the tree to question her), visited by supportive celebrities, and considered an inspiration to other activists.

In December 1999 an agreement was reached with Pacific Lumber Company that created a 200-foot buffer zone around the tree, and assurances were given that Luna would not be cut. Hill and her supporters agreed to pay $50,000, which the company pledged to forestry research at Humboldt State University in nearby Arcata, California.

Since her record tree-sit, Julia Hill has become an environmental celebrity. Her book, *The Legacy of Luna*, recounts her experiences as a protester, and she coauthored a second book, *One Makes the Difference*. She has formed a nonprofit organization, Circle of Life, to promote sustainability, spoken on college campuses and at high schools, and has become involved in numerous other protests on environmental and social justice issues.

Since then, the tree at the center of the protest has been vandalized with a chainsaw that made deep cuts into its core, although no individual or group has claimed responsibility.

# Andy Kerr (b. 1955)

Creswell, Oregon, referred to by Andy Kerr as a "recovered timber town," is located in Oregon's Willamette Valley. Large logging-industry towns like Medford and Roseburg may be better known, but in the upper Willamette, where he was born, the timber industry was the heart of the town. A fifth-generation Oregonian, Kerr's migration away from the Creswell norm began in high school, where at age seventeen he joined the group Zero Population Growth. He attended Oregon State University in Corvallis, studying forestry, political science, economics, history, and, as he puts it, "beer," but did not graduate.

In 1976 he started working with the Oregon Natural Resources Council, one of the state's strongest and most influential environmental advocacy groups. He rose to the position of executive director and spent two decades embroiled in the timber wars that typified the Pacific Northwest in the 1980s and 1990s. During his tenure with the organization, the region's timber companies battled environmental groups over the fate of the Northern spotted owl, a bird that was eventually declared a threatened species. He was invited to participate in the 1993 Northwest Forest Conference in Portland, convened by President Bill Clinton in an effort to bring the various stakeholders to the table.

He now serves in a triad of positions: as director of the National Public Lands Grazing Campaign (NPLGC), as a board member and treasurer of the North American Industrial Help Council, and as self-described "czar" of his own firm, the Larch Company, which is named after a coniferous tree. He has also served as a consultant for the Wilderness Society and as an adviser to the Soda Mountain Wilderness Council, and has a lengthy list of venues in which he has lectured or been interviewed. Kerr is the author of two books, *Oregon Desert Guide: 70 Hikes* and *Oregon Wild: Endangered Forest Wilderness.*

Kerr also is recognized for his dedication to natural resource issues, and he has been given awards and accolades from a variety of sources. He has been compared to Ralph Nader in his tireless work to preserve old growth forests. One *New York Times* reporter notes: "He has a talent for speaking in such loaded sound bites that it was said by reporters that if Andy Kerr did not exist, someone would have to invent him."

In his position with the NPLGC, Kerr has now targeted subsidies given to ranchers that allow them to graze livestock on public lands at below-market prices. Kerr has proposed a government buyout program that would pay ranchers to stop grazing cattle in exchange for "generous" compensation. Although the proposed buyout has been criticized for being too costly, Kerr argues that the long-term expense is outweighed by the environmental benefits to fish, wildlife, watersheds, and recreation. "I'm not conflict averse," he once said. "Nobody got anywhere by being nice."

# J. Horace McFarland (1859–1948)

Gifford Pinchot is perhaps the best-known natural resource leader to hail from Pennsylvania, and he has often overshadowed another native son whose work was also important in the development of national parks and protected areas.

J. Horace McFarland was born in McAlisterville, Pennsylvania, on September 29, 1859, and has often been called the forgotten hero of the preservation movement. His first name was John, but he always went by Horace. His parents operated the McAlisterville Academy, where he was born. His father, Lt. Col. George F. McFarland, was severely wounded during the Civil War at the battle of Gettysburg, but he survived and moved the family to Harrisburg in 1865. McFarland went to private schools from 1867 to 1871, but, unlike Pinchot, that was the only formal schooling he ever received.

His father purchased the Riverside Nurseries Company in Harrisburg, and his son quickly learned the horticulture business, becoming a master gardener. McFarland established his printing business, J. Horace McFarland Co., in 1878, which was known for its progressive work environment and policies. The plant was clean and workers' health concerns were important; he offered a benefits package to his employees that included vacations, insurance, and assistance in purchasing a home, and he assigned women management responsibilities in his business.

McFarland was a strong believer in civic duty, and worked for forty-five years with the Municipal League of Harrisburg. He also served as president of the American League for Civic Improvement, the American Park and Outdoor Art Association, and as a founding leader of the American Civic Association. His "Harrisburg Plan" became a model of municipal reform, emphasizing clean water, sewage systems, and the preservation of city parks. He traveled across the United States, crusading against civic blight and encouraging citizens to write to Congress about environmental issues and what would become the City Beautiful movement.

From 1909 until 1913, McFarland joined with John Muir and the Sierra Club in the battle to prevent the building of a dam in Yosemite's Hetch Hetchy Valley. They were unsuccessful, and while Muir retreated from the defeat, McFarland contacted two of the men who had championed the Hetch Hetchy legislation about an idea he had been considering. While in Washington, D.C., McFarland had discovered that there was not "a single desk" in the city devoted to the administration of the nation's parks, and he was among the first to call for a unified federal agency to manage monuments and parks. His idea gained support from San Francisco's city attorney and from the congressman who had persuaded President Woodrow Wilson to sign the Hetch Hetchy legislation. As a result of his efforts, the National Park Service was established in 1916. In 1920 he joined the fight to save Yellowstone Lake, and in 1924 he received an honorary degree from Dickinson College. McFarland served on the Department of the Interior's Educational Advisory Board and as a member of the National Park Trust Fund until his death.

# Chico Mendes (1944–1988)

Chico Mendes was born in a small village in Brazil's western Amazon rain forest, the son of a poor family of rubber tappers, called *seringueiros*. Although he was illiterate during his childhood, he would become one of the first guardians of the Amazon forests, helping to bring an end to governmental grazing subsidies that had caused the destruction of thousands of acres of land in the fragile ecosystem.

At age nine, Mendes joined his family in the tradition of generations of poor Brazilians who made a meager living collecting the white, milky latex from the trees in the Amazon. At the time, landowners did not allow their workers to build or attend schools, so his education was an informal one. A friend who was a political refugee taught him how to read and write, using old magazines and a short-wave radio.

In the 1970s, Mendes became a leader in a nonviolent resistance movement to defend their homes from cattle ranchers, who demanded that they leave. Along the western border of Brazil the government had begun its National Integration Program, hoping to colonize the region with cattle ranchers and forcing the native people to relocate. Over the next fifteen years the ancient forests were intentionally burned to make way for farms and ranches, resulting in massive erosion and a loss of jobs. Mendes organized his fellow rubber tappers to protest the relocation, organizing blockades against bulldozers. He founded a trade union in Acre, Brazil, in 1975 and the Workers' Party four years later. In 1985 he expanded the movement by creating the National Council of Rubber Tappers. He sought the assistance of environmental groups in the United States to build rubber tree preserves that would provide the local people with a source of income by practicing sustainable agriculture. He built a coalition between the rubber tappers and the indigenous peoples, and his leadership and power became a threat to the local ranchers.

Throughout 1988, Mendes and his family received numerous death threats. He was murdered just outside his home while his wife and children watched nearby. Guards who had been hired by the family for protection had suddenly disappeared, and it was not until the international environmental community mobilized that the Brazilian government began to investigate his assassination. Two local ranchers were eventually arrested and charged with Mendes's death; his wife became president of the Chico Mendes Foundation.

# Milford Muskett (b. 1967)

Life in a rural community often influences those who work in the field of natural resources. For Milford Muskett, who was born and raised on the Navajo Reservation, rural is perhaps an understatement. The nearest town was Tohatchi, New Mexico, hardly a familiar name even to those who know the U.S. Southwest. However, Muskett says that life on the reservation had very little influence on his desire to go into the field of environmental research and teaching. "I was drawn to the environmental field through stories from Aldo Leopold," the co-founder of the Wilderness Society and author of *The Sand County Almanac.*

He left the reservation to pursue an education, receiving a bachelor's degree in biology and geography at Calvin College in Grand Rapids, Michigan, in 1990. From there he moved to Western Michigan State University at Kalamazoo, obtaining a master's degree in geography in 1993. Moving to the University of Oklahoma, he attended several other graduate programs before earning his Ph.D. in land resources in 2003 at the Gaylord Nelson Institute for Environmental Studies at the University of Wisconsin, Madison. The institute is named for Senator Gaylord Nelson, who is considered the "father" of the first Earth Day on April 22, 1970.

While in Wisconsin, Muskett studied under the tutelage of historian William Cronon, who became his academic adviser. "His work and his mentorship have helped me understand the need for storytelling and the need for creating good history about the environment and people interaction," he says. Although he had started his Ph.D. work with an emphasis in conservation biology, at Wisconsin he shifted gears to Indian environmental history. He taught history at Marquette University and joined the faculty at Cornell University as a visiting assistant professor in 2004, teaching in the Department of Natural Resources and the American Indian Studies program. He also served as an adviser to the American Indian Science and Engineering Society. His research interest is the conflicts between the cultural and political aspects of traditional societies and Western societies around natural resources development and environmental regulations. More specifically, he examines how storytelling and traditional ecological knowledge are used to define and teach ecological relationships. On a personal level, he likes to bake and calls himself "a Jack of All Trades."

Muskett's career has a pragmatic policy side as well. He has worked with the Navajo Nation's Environmental Protection Agency, which motivated him to understand and learn more about environmental issues on the reservation, and also worked as a cartographer for the International Crane Foundation. He has been selected as a fellow of the Environmental Leadership Program's (ELP) Class of 2006–2007. Each year, ELP selects about twenty top environmentalists from across the nation to participate in a two-year program that supports talented emerging leaders from academia, government, business, and nonprofit organizations through retreats, networking, and training opportunities.

# Joan Norman (1933–2005)

"I would rather go out in a blaze, defending the world I love. I will be on the front lines someday and my soul will know the time to go and I will just leave." Activist Joan Norman made those comments in March 2005, just months before she was killed in a head-on car collision on Highway 199 near the California border. Her death, on July 23, 2005, stunned forest defense advocates who had depended on Norman for her contagious resolve and humble nobility, which challenged those around her to take a stand for what they hold most dear.

Her comments appeared in newspapers around the United States, along with a photograph of Norman seated in a walker and with a cane on the Green Bridge in southern Oregon, beneath a U.S. flag and a banner protesting the Biscuit Fire Recovery Project. She was arrested twice that month and voluntarily spent several weeks in jail in protest of illegal logging, refusing to post bail. While in jail she helped other inmates by offering legal support, even though she had never before had a lawyer herself. What makes Joan Norman stand out among forest activists is that when she was arrested, she was seventy-two years old.

Norman was born in Oklahoma in 1933 into what she called "a culture that trashed the earth, enslaved the earth to extract wealth." She had come from a family of Republicans, and she herself married into wealth. "I did my wifely duties so that we could keep our money." She was inspired when John F. Kennedy, Jr., ran for president in 1960 because she felt that he spoke directly to the people. The stereotype of antilogging activists is a young man with long hair or dreadlocks, wearing a tie-dyed shirt, torn jeans, and sandals, chained to an old growth redwood tree. Norman represented an entirely different generation, arrested more than 100 times during her life for various civil rights and environmental issues. She was not a newcomer to civil disobedience or controversy when she was arrested in Oregon.

She began her civil rights activism in California, joining members of a church headed to Alabama. "I walked with Martin Luther King, Jr. The thing we wanted to stand up to then was the destruction of the diversity of people in this nation. The slavery, racism, and violence toward people of color." She joined in the growing anti–Vietnam War movement, later participating in demonstrations at the School of the Americas in Ft. Benning,

Georgia. She lived in a motor home for twelve years, traveling from one demonstration to another. Norman joined members of the Western Shoshone tribe in an effort to stop the mining of uranium in New Mexico. Her journey led her to Seattle and Washington, D.C., to protest against the World Trade Organization. "I had my own kitchen, my own first aid station, my few books and my passion for freedom and justice."

Norman had gone from Selma, Alabama, to Selma, Oregon, joining the Siskiyou Forest Defenders in a peaceful resistance campaign to protest logging of timber burned in the Biscuit Fire. The U.S. Forest Service had authorized a logging operation encompassing about 20,000 acres, with a goal of 372 million board feet—the largest in the agency's history. She sat in her walker on a bridge used by the Silver Creek Logging Company, trying to block access to what protesters said was an illegal old growth logging sale on Fiddler Mountain. She was joined by twenty other women wearing black, "in solidarity with the trees." In August 2005, protesters joined together in the Joan Norman Memorial Road Blockade near the site of the Hobson old growth timber sale in remembrance of their friend.

# William Penn (1644–1718)

The son of an English admiral, William Penn was born in England and grew up in the Essex countryside, attending the Puritan Chigwell Grammar School, whose theological foundations would later affect his attitudes toward religious toleration. In 1655 the family moved back to London and then to Ireland, and in 1660 he entered the University of Oxford. He was expelled in 1662, not for his academic work but for his religious beliefs. He had heard a Quaker, Thomas Loe, preach while in Ireland and had rejected the dominant Anglican faith. To put his son back on the traditional track, Admiral Penn sent William on a European tour, and afterward, he entered the Protestant College at Saumur, France, far from the religious upheaval he had become involved with in Great Britain. He then returned to England and spent a year reading law before his father sent him back to Ireland in 1666 to manage the family's estates.

He is less well known for his influence on early American conservationism. He served as a trustee for one of the two

Quaker proprietors of West New Jersey, and in 1681, Penn and eleven other Quakers purchased East New Jersey. At age thirty-seven, he was also given land on the west bank of the Delaware River as payment for a large debt owed by Charles II, and he called the province Pennsylvania (Penn's woods) after his father, who had died in 1670. He sought to create a refuge for his fellow Quakers, who were being persecuted, creating what he called "a holy experiment."

As the colony's new governor, Penn came to North America and established treaties with the Delaware Indians based on mutual trust. The experiment included an elaborate plan for the growth of a new city, Philadelphia, which was laid out in a grid pattern surrounded by farming land. He interpreted the name as "the city of brotherly love" but also referred to it as Greene Country Towne. By 1700 the settlement had become the second largest city in the New World, but Penn preferred to stay at his wilderness home, Pennsbury, in the thick woods across the river from what is now Trenton, New Jersey. He loathed the crowded, dirty cities of Europe, and told his surveyors to make sure that the town was made up of evenly spaced streets with wide boulevards. He insisted that there be five public squares of greenery, and that each plot of land within the town was at least an acre, with room for gardens, orchards, or fields.

He wrote: "Here, the air is sweet and clear, the heavens serene. The woods are adorned with lovely flowers for colour and variety. I have seen the gardens of London with that sort of beauty, but think they may be improved by our woods." To make sure that the woods survived the growing population, he decreed that for every tree cut, five would be left untouched—one of the colony's first environmental laws. Pennsylvania was to be governed by stewards of the lands, even though the colony was exporting large amounts of timber, furs, hemp, tobacco, iron, and copper in exchange for British goods.

Penn made only two visits to North America; from 1682 to 1684 he focused on boundary disputes with Lord Baltimore, and during his second trip, from 1899 to 1701, he met with officials to establish an internal government and the adoption of a new form of government under the Charter of Privileges. Returning to England, he believed that the stewardship concept would protect the resources of the New World. In 1712 he suffered a severe stroke, and he died at the age of seventy-three in 1718. He never really had an opportunity to enjoy Pennsbury or his deep woods, but he set a precedent for the protection of the land that would be followed by other colonies through the time of the Revolutionary War.

# Sandra Postel (b. 1956)

Historically, the field of natural resources research has been dominated by men, with a few women in government agencies, academia, or nongovernmental organizations. Sandra Postel, director of the Global Water Policy Project in Amherst, Massachusetts, is notable for both the breadth of her work and research and the contributions she has made to understanding key conservation problems and solutions. She is considered one of the world's leading authorities on international water policies and sustainability.

She grew up on Long Island, New York, and her academic training began at Wittenberg University, where she received her bachelor of arts degree in geology and political science. In 1980 she was awarded a master's degree from Duke University with an emphasis on resource economics and policy, and she has also received two honorary doctor of science degrees. She has an extensive list of both scholarly and popular articles, ranging from op-ed pieces in the *New York Times* to *Scientific American*, and appears on radio and television programs as a commentator and analyst. Postel's awards are evidence of her continuing role in the field; she was named one of the *Scientific American* 50 in 2002 for her contributions to science and technology, and for promoting sweeping changes aimed at preserving the world's dwindling supplies of fresh water. In addition to serving as a member of several editorial boards and boards of directors, she is an adviser to the Division on Earth and Life Studies of the U.S. National Research Council. She has been awarded the Duke University School of Environment's Distinguished Alumni Award, a Pew Scholars Award in Conservation and the Environment (1995), and a lifetime chair with the International Water Academy in Oslo, Norway. Prior to founding the Global Water Policy Project, she served as vice president for research for the Worldwatch Institute from 1988 to 1994, and she continues her research there as a senior fellow. In *Pillar of Sand*, she calls water scarcity the single most important threat to global food production. As the population increases, and the world's supply of fresh water decreases, there is both a need for better water allocation and a need to make irrigation more effective. Her most recent book, *Rivers for Life: Managing Water for People and Nature,* explores the disruption of natural river flows by more than 45,000 large dams blocking the world's rivers. Postel explains why restoring and preserving more natural river flows are key to sustaining freshwater biodiversity and healthy river systems.

By emphasizing the finite nature of water, and humanity's efforts to control it and manage it as a natural resource, Postel sounds an important alarm for the world's policy-makers. But she also believes that today's youth will bring a heightened awareness and appreciation of the environment to the generations that follow. In addition to her expertise and ability to understand scientific processes, she brings optimism to water resource issues at a time when scarcity is more serious than ever.

# Ken Saro-Wiwa (1941–1995)

There is a historical tradition that many of the world's most influential environmental activists began as writers and poets, as is the case with Nigerian author Ken Saro-Wiwa. He was born Kenule Benson Tsaro-Wiwa in Bori, Rivers State, and was considered a child prodigy. At age thirteen he was awarded a scholarship to Government College in Umuahia, Nigeria, and later graduated from the University of Ibadan. In 1985 he published his first novel, the first of more than fifty he would write during his lifetime. His legacy includes a children's television series, radio plays, political columns, and essays.

How does a writer become an environmental activist? In 1958, the oil company Royal Dutch Shell began drilling in the Niger Delta on coastal land inhabited by more than a half-million Ogoni people. Over the years the fertile farmland was turned into an oil dump, killing wildlife and fish, and ruining the lives of the farmers and fishers. The area was Saro-Wiwa's homeland, and he was outraged at the human and environmental cost to the region. More than 5,000 workers had been hired by Shell, but fewer than 100 were Ogoni. Almost none of the revenue from the oil operations had trickled back down to the local residents, and the company refused to take any responsibility for the cleanup when it stopped drilling in 1993.

In 1990, Saro-Wiwa founded the Movement for the Survival of the Ogoni People (MOSOP); he was also alleged to have started a radical youth group engaged in sabotage against the oil company. In 1991–1992 he wrote two books that criticized the corruption of the Nigerian government in its relations with Shell Oil and British Petroleum, laying the blame for the environmental damage on the British government. He and his supporters sought compensation for the Ogoni people and intervention to clean up the environmental damage, actions that brought threats and intimidation from the Nigerian government and military leaders. On January 4, 1993, Saro-Wiwa organized more than 300,000 Ogoni, who marched in protest of the government's complicity in the oil pollution scandal; the date is now celebrated as Ogoni Day.

Nigerian military forces began attacking Ogoni villages, killing thousands of people and leaving villagers homeless. In May 1994, Saro-Wiwa was taken from his home and arrested, along with eight other activists, and charged with the murder of four Ogoni leaders. The human rights group Amnesty International named him a prisoner of conscience, and other groups complained that the arrests were based on trumped up charges. After a brief show trial, Saro-Wiwa and his codefendants were hanged by the military on November 10, 1995. Shell Oil made no official comment on the execution.

# Maurice Strong (b. 1929)

Often there are unsung leaders whose behind-the-scenes efforts are overshadowed by those in positions of power, even though they have worked tirelessly to resolve conflicts. Such is the case with Maurice Strong of Canada, who is less known than others who are credited with the world's first Earth Summit in 1972.

Strong was born in 1929 and became determined to improve his fortune despite his childhood in a poor family in rural Manitoba. He completed his secondary education and ran away from his family at age fourteen to join the Canadian merchant marine. His father found him in Vancouver, British Columbia, and he soon left home again to become an apprentice to a fur trader in the Arctic. His new family would consist of the Inuit, the indigenous people of the region. In 1947 he turned eighteen years old and met a UN official at a dinner party who helped him to get a position with the UN Security Department in New York. He left the job after two months, although he was able to make contacts with many prominent leaders. He tried to join the Canadian air force but failed to qualify; he then obtained employment with Dome Petroleum as an analyst.

In another job shift, he left the company and took a world cruise, ending up in Nairobi, Kenya, where he worked for another oil company looking for sites for new gas stations in Africa. As had been the case in the Arctic, Strong gained close ties with the native people and began working with local organizations until he returned to Canada and Dome Petroleum in 1955. From there he leapfrogged up the corporate ladder to the Power Corporation of Canada, and received a political appointment as director general of Canada's International Development Agency. The UN secretary-general, U Thant, met Strong and asked him to organize the world's first international environmental conference. From November 1970 until December 1972, Strong served as the secretary-general of the UN Conference on the Human Environment in Stockholm, Sweden. He moved to Geneva to make arrangements for the meeting and was met by leftover Cold War resistance. Many communist leaders refused to participate, although he was able to convince China of the benefits of attending. He also bridged the gap between developed and developing countries that had not yet been given a seat at the international table in environmental matters. A total of 113 UN members sent delegates, although the Soviet Union refused to participate and most heads of state did not attend. Still, the Stockholm Declaration produced by the delegates was the first international agreement on environmental goals, including acceptance that clean air and water are human rights.

Strong was named the first director of the UN Environment Programme, based in Nairobi, where he served from January 1973 to December 1975. He returned to Canada to serve as the head of the country's national oil company, and from 1985 to 1986 he returned to the United Nations as undersecretary-general and as a member of the World Commission on Environment and Development. Twenty years after Stockholm, he served as secretary-general of the UN Conference on Environment and Development, also known as the Earth Summit, in Rio de Janeiro. There, the issue of sustainability took center stage and nongovernmental organizations became major participants for the first time.

After the conference, Strong was named chief executive officer and chairman of Ontario Hydro, the largest utility company in North America, where he worked from 1992 to 1995. He joined the World Bank as senior adviser to the president in June 1995. He returned to the United Nations again to serve as the envoy to North Korea, where he became embroiled in scandal over allegedly serving as an unregistered agent for the Iraqi government. He stepped down from his UN position in April 2005 as the investigation continued.

Strong has many supporters as well as many detractors, largely because of the many career changes he has gone through, switching back and forth between the private sector and public service. But even those who oppose some of his policies admit that he was one of the world's most influential environmental leaders, even when he was backstage, riding a bicycle painted in the UN's characteristic blue and white colors.

# Nicola Temple (b. 1973)

"In my very limited experience in environmentalism, I would say there are two types: the environmentalist who turns to science to provide fact for their arguments and the scientist who turns to environmentalism because what they set out to study is threatened by human activities. I would say that I fall into the latter category."

How do environmentalists get started in their careers? How do they feel they can make a difference? In British Columbia, where controversies over timber harvests have taken center stage, a young Canadian woman, Nicola Temple, got her start at a young age.

She grew up in the Ottawa Valley in Canada, where her family grew all of their own food, and she spent hours exploring under logs, digging around in ponds, and canoeing the local rivers. She obtained a degree in biology from the University of Victoria in 1998 and then took time off to travel. She returned to UVic and received a master's degree in 2003, researching functional morphology. Her specialization was the function of the adipose fin in salmon and the hydrodynamic implications of its removal, a common practice in fish hatcheries.

While working on her master's degree, she met graduate students who were approaching the environment from a different perspective. "They were researching various aspects of how salmon enrich and increase diversity in an area by bringing marine-derived nutrients to terrestrial ecosystems when they return to their natal rivers to spawn and then die. This was my 'aha' moment with respect to the connectivity of ecosystems," she says.

Seeking work that would be meaningful and would translate directly into the improved management of resources, she applied for a position with the Raincoast Conservation Society in Victoria, where she has worked since 2004. Her responsibilities often involve working in remote areas along British Columbia's coast, being on the ground and conducting fieldwork, and working closely with coastal communities. Temple also meets with other environmental group leaders and governmental officials to improve policy options. She now works on studying salmon populations in small streams throughout the coast, which, when combined, contribute substantially to the fishery and the number of fish returning to coastal systems annually. The small salmon populations are also extremely important to the local communities that depend upon them for food and cultural purposes.

"Though being in the midst of [the salmon] debate and controversy often wears on the emotions, I can't imagine being anywhere else. Salmon need a voice . . . many voices. Perhaps it is my naivete, but I think we can solve this issue."

# Courtney White (b. 1960)

The tagline "Restoration, innovation, and education, one acre at a time" describes the work of the Quivira Coalition, based in Santa Fe, New Mexico. The group's executive director, Courtney White, is part of the "New Ranch" movement in the western United States, which seeks to save ranching operations while employing tools such as stream restoration, high-intensity rotational grazing, and collaboration.

White did not grow up on a ranch, and even now he lives in the suburbs rather than in cattle country. Growing up in Phoenix, he graduated from Reed College, a small liberal arts school in Portland, Oregon, as an anthropology major. He decided to go to film school at the University of California, Los Angeles, and then worked as an archaeologist for the National Park Service in Santa Fe. White says that his interest in environmental activism began in 1994 with the election of a Republican majority in Congress, and Newt Gingrich's Contract with America. "I couldn't be a passive observer any more," he says, and so he contacted the local chapter of the Sierra Club. He began by serving as a volunteer tracking legislation, and became a member of the executive committee in the summer of 1995.

As part of his work with the statewide Sierra Club, White met rancher Jim Winder, who gave him a tour of his ranch near Deming, New Mexico, in January 1996. Winder had grown disillusioned with the Sierra Club; he and White started a nonprofit "alternative" group, the Quivira Coalition, with a $1,000 donation from Winder as seed funding. The name comes from the Spanish term used to describe uncharted territory.

As executive director of the coalition, White works with governmental agencies such as the U.S. Forest Service and the New Mexico Environment Department, along with university researchers and ecologists. He is best known for bringing together the broad range of grassroots organizations that represent stakeholders in grazing controversies, creating in 2003 the Southwest Grassfed Livestock Alliance, which seeks to educate consumers about the benefits of eating beef raised on the open range. As part of the "radical center," White and others have called for a truce among those who have been arguing over grazing for decades.

He believes that sustainable ranching is possible in the West, contending that there is sufficient scientific study available to show that grazing can be done in a rest-rotation pattern that allows vegetation to regrow. As part of a shift toward progressive cattle management, White believes that the Quivira Coalition provides a neutral place for discussion and the exchange of ideas. The group now focuses its efforts on large landscape restoration, and White plans to continue his work as an active participant in the rangeland debate.

# Terry Tempest Williams (b. 1955)

"I write through my biases of gender, geography, and culture. I am a woman whose ideas have been shaped by the Great Basin and Colorado Plateau." As one of the nation's foremost nature writers, Terry Tempest Williams has played an important role in awakening Americans to the need to protect wild places. In 1991, *Newsweek* named her someone likely to make "a considerable impact on the political, economic, and environmental issues facing the western states this decade."

Such an accolade might seem improbable for a woman whose family was part of the original Mormon migration from Nauvoo, Illinois, to arrive in Salt Lake City, Utah, in 1846. She had a traditional Mormon upbringing, living with several generations of family members, and at age seventeen she had a religious vision that she says confirmed her faith in the Church of Jesus Christ of Latter-Day Saints. Although a believer, however, she would also develop a rebellious streak that caused her to declare that she was not an "orthodox Mormon."

Where do concerns about the environment begin in someone's life? For Williams, it was the trips that she took as a child with her maternal grandmother to the Bear River Migratory Bird Refuge. She grew up in Salt Lake City, and the visits to the open land of Utah's deserts led her to pursue a dual degree in biology and English at the University of Utah. She also received a master's degree in education with a specialization in environmental education. She worked on the Navajo Reservation while working on her graduate degree, and wrote her first set of essays, *Pieces of White Shell*, in 1984. She had intended the work to be her master's thesis, but her graduate committee rejected it. "That is," she says, "until Scribner accepted it for publication. Then they reversed their decision." In 2003 she received an honorary doctorate from the University of Utah, and a similar award was made by Saint Mary-of-the-Woods College in 2004.

Williams has been honored in numerous ways for her tenacious dedication to preserving the West. She has been included in the Rachel Carson Honor Roll, has received the National Wildlife Federation's Conservation Award for Special Achievement, was named one of the *Utne Reader*'s Utne 100 Visionaries, was named a fellow of the John Simon Guggenheim Memorial Foundation, and received a Lannan Literary Fellowship in Creative Non-Fiction. She was also honored as the Annie Clark Tanner Scholar at the University of Utah and recipient of the 2005 Wallace Stegner Award by the Center for the American West.

While she may be best known for her writing, Williams has also been an environmental activist. She organized twenty other writers to contribute their essays on why it is important to protect wildlands. *Testimony: Writers Speak on Behalf of Utah Wilderness*, which she edited and was published in 1996, had an impact that reached all the way to the White House. When President Bill Clinton dedicated the Grand Staircase-Escalante National Monument on September 18, 1996, he held up a copy of the book at the ceremony on the North Rim of the Grand Canyon. "This," he said, "made a difference."

# Howard Zahniser (1906–1964)

The background of one of the great leaders of the U.S. wilderness preservation movement, Howard Zahniser, provides a handful of clues to his professional activism and interests. He was born in Pennsylvania to a minister whose wife was a Seneca tribe descendant, and he joined the Audubon Society in the fifth grade. As editor of the student newspaper at Greenville College, where he received a degree in English literature, his skills translated into a career in journalism and as a high school teacher. Later he would work as an editorial assistant with the agency that would become the U.S. Fish and Wildlife Service.

At age twenty-four his career in the federal government was typical, as he transferred from one agency to another. Along the way he met a series of influential writers, biologists, and naturalists who uniquely molded his view of the world. He worked briefly with Rachel Carson and with Aldo Leopold; he was mentored by Edward Preble and J. "Ding" Darling. He joined the Wilderness Society, which had been founded in 1935 by Robert Marshall, as a charter member. By 1942, his expanding interests in natural resource management led to a position as the publicist for the government's Victory Gardens campaign during World War II and as a writer for several nature-oriented publications.

Marshall died in 1939, and the Wilderness Society's new director, Olaus Murie, died in 1945. Zahniser was asked to serve as executive secretary of the organization, and he assumed responsibility for its magazine, *The Living Wilderness;* he also wrote a current events section. Along with David Brower, director of the Sierra Club, he led a coalition of environmental groups in the 1950s in successfully opposing the building of Echo Park Dam in Colorado's Dinosaur National Monument. Their combined activism led to federal legislation which provided that no dam or reservoir would subsequently be constructed within a national park or monument.

He is most remembered, though, for his leadership in advocating for national wilderness legislation: between 1956 and 1964, Zahniser wrote sixty-six drafts of a bill, and steered the proposals through eighteen congressional hearings. As a writer, he sought to convince the members of Congress that certain pristine lands needed protection, but wanted language that would convey just what kind of land was being discussed. A friend remarked that she enjoyed the "untrammeled" seashores of Olympic National Park, and Zahniser had found the word he wanted.

After years of failing health, he died in May 1964 at age fifty-eight at his home in Maryland. Just four months later his wife, Alice, was an honored guest in the White House Rose Garden when President Lyndon Johnson signed the Wilderness Act that Zahniser had fought for. It established the 9-million-acre National Wilderness Preservation System, giving Congress the power to recommend and designate future lands that would be added. In 1998, Greenville College dedicated the Zahniser Institute for Environmental Studies in honor of its most distinguished alumnus.

# 6

# Documents and Data

## Legislation

In the United States, legislation enacted by Congress provides a revealing look at the intent of legislators dealing with natural resource conflicts. In most cases, the statutes passed have provided protection, such as for national parks, but in some instances legislation has been designed to advance the agenda of extractive resource industries, such as fossil fuels or timber. This section contrasts approaches by providing the text or excerpts from some of the key public laws over which conflicts have occurred.

Considered one of the most important statutes in the development of the U.S. park system, the National Park Service Organic Act, signed in 1916, brought together for the first time the independent units of protected areas that had been under three cabinet agencies. The legislation passed by Congress recognized the importance of scenic areas to the cultural heritage of the United States, as well as the need for official protection to preserve identified sites. In contrast, the Taylor Grazing Act, enacted in 1934, exemplifies the concept of multiple use, as Congress provided a mechanism for allowing ranchers to graze their livestock on public lands. A third key statute, the Federal Land Policy and Management Act of 1976 (FLPMA), provides a more contemporary statement of congressional intent toward management of the public lands, incorporating the issues of disposal and withdrawal, public participation in natural resource policy-making, the protection of values as defined in the law, and conflicting uses for mining, recreation, wildlife, and human occupancy.

## National Park Service Organic Act

Be it enacted by the Senate and House of Representatives of the United States of America in Congress assembled, That there is hereby created in the Department of the Interior a service to be called the National Park Service, which shall be under the charge of a director, who shall be appointed by the Secretary and who shall receive a salary of $4,500 per annum. There shall also be appointed by the Secretary the following assistants and other employees at the salaries designated: One assistant director, at $2,500 per annum; one chief clerk, at $2,000 per annum; one draftsman, at $1,800 per annum; one messenger, at $600 per annum; and in addition thereto, such other employees as the Secretary of the Interior shall deem necessary: Provided, That not more than $8,100 annually shall be expended for salaries of experts, assistants, and employees within the District of Columbia not herein specifically enumerated unless previously authorized by law. The service thus established shall promote and regulate the use of the Federal areas known as national parks, monuments, and reservations hereinafter specified by such means and measures as conform to the fundamental purposes of the said parks, monuments, and reservations, which purpose is to conserve the scenery and the natural and historic objects and the wildlife therein and to provide for the enjoyment of the same in such manner and by such means as will leave them unimpaired for the enjoyment of future generations.

SEC. 2. That the director shall, under the direction of the Secretary of the Interior, have the supervision, management, and control of the several national parks and national monuments which are now under the jurisdiction of the Department of the Interior, and of the Hot Springs Reservation in the State of Arkansas, and of such other national parks and reservations of like character as may be hereafter created by Congress: Provided, That in the supervision, management, and control of national monuments contiguous to national forests the Secretary of Agriculture may cooperate with said National Park Service to such extent as may be requested by the Secretary of the Interior.

SEC. 3. That the Secretary of the Interior shall make and publish such rules and regulations as he may deem necessary or proper for the use and management of the parks, monuments, and reservations under the jurisdiction of the National Park Service, and any violations of any of the rules or regulations authorized by this Act shall be punished as provided for in section 50 of the Act entitled, "An Act to codify and amend the penal laws of the United States," approved March fourth, nineteen hundred and nine, as amended by section six of the Act of June twenty-fifth, nineteen hundred and ten (Thirty-sixth United States Statutes at Large, page eight hundred and fifty-seven). He may also, upon terms and conditions to be fixed by him, sell or dispose of timber in those cases where in his judgment the cutting of such timber is required in order to control the attacks of insects or diseases or otherwise conserve the scenery or the natural or historic objects in any such park, monument, or reservation. He may also provide in his discretion for the destruction of such animals and of such plant life as may be detrimental to the use of any of said parks, monuments, or reservations. He may also grant privileges, leases, and permits for the use of land for the accommodation of visitors in the various parks, monuments, or other reservations herein provided for, but for periods not exceeding thirty years; and no natural curiosities, wonders, or objects of interest shall be leased, rented, or granted to anyone on such terms as to interfere with free access to them by the public; Provided, however, That the Secretary of the Interior may, under such rules and regulations and on such terms as he may prescribe, grant the privilege to graze live stock within any national park, monument, or reservation herein referred to when in his judgment such use is not detrimental to the primary purpose for which such park, monument, or reservation was created, except that this provision shall not apply to the Yellowstone National Park: And provided further, That the Secretary of the Interior may grant said privileges, leases, and permits and enter into contracts relating to the same with responsible persons, firms, or corporations without advertising and without securing competitive bids: And provided further, That no contract, lease, permit or privilege granted shall be assigned or transferred by such grantees, permitees, or licensees, without the approval of the Secretary of the Interior first obtained in writing: And provided further, That the Secretary may, in his discretion, authorize such grantees, permitees, or licensees to execute mortgages or bonds, shares of stock, and other evidences of interest in or indebtedness upon their rights, properties, or franchises, for the purposes of installing, enlarging or improving plant and equipment and extending facilities for the accommodation of the public within such national parks and monuments.

SEC. 4. That nothing in this Act contained shall affect or

modify the provisions of the Act approved February fifteenth, nineteen hundred and one, entitled "An Act relating to rights of way through certain parks, reservations, and other public lands."

# Excerpts, Taylor Grazing Act of 1934

### Title 43, Chapter 8A, Subchapter I

Section 315. In order to promote the highest use of the public lands pending its final disposal, the Secretary of the Interior is authorized, in his discretion, by order to establish grazing districts or additions thereto and/or to modify the boundaries thereof, of vacant, unappropriated, and unreserved lands from any part of the public domain of the United States (exclusive of Alaska), which are not in national forests, national parks and monuments, Indian reservations, revested Oregon and California Railroad grant lands, or revested Coos Bay Wagon Road grant lands, and which in his opinion are chiefly valuable for grazing and raising forage crops; Provided That no lands withdrawn or reserved for any other purpose shall be included in any such district except with the approval of the head of the department having jurisdiction thereof. Nothing in this subchapter shall be construed in any way to diminish, restrict, or impair any right which has been heretofore or may be hereafter initiated under existing law validly affecting the public lands, and which is maintained pursuant to such law except as otherwise expressly provided in this subchapter nor to affect any land heretofore or hereafter surveyed which, except for the provisions of this subchapter, would be a part of any grant to any State, nor as limiting or restricting the power or authority of any State as to matters within its jurisdiction.

Whenever such grazing district is established pursuant to this subchapter, the Secretary shall grant to owners of land adjacent to such district, upon application of any such owner, such rights-of-way over the lands included in such district for stock-driving purposes as may be necessary for the convenient access by any such owner to marketing facilities or to lands not within such district owned by such person or upon which such person has stock-grazing rights . . .

Before grazing districts are created in any State as herein provided, a hearing shall be held in the State, after public notice thereof shall have been given, at such location convenient for the attendance of State officials, and the settlers, residents, and livestock owners of the vicinity as may be determined by the Secretary of the Interior. No such district shall be established until the expiration of ninety days after such notice shall have been given, nor until twenty days after such hearing shall be held. . . . Nothing in this subchapter shall be construed as in any way altering or restricting the right to hunt or fish within a grazing district in accordance with the laws of the United States or any State, or as vesting in any permitee any right whatsoever to interfere with hunting or fishing within a grazing district.

Section 315A. The Secretary of the Interior shall make provision for the protection, administration, regulation, and improvement of such grazing districts as may be created under the authority of Section 315 of this title, and he shall establish such service, enter into cooperative agreements, and do any and all things necessary to accomplish the purposes of this subchapter and to insure the objects of such grazing districts, namely, to regulate their occupancy and use, to preserve the land and its resources from destruction or unnecessary injury, to provide for the orderly use, improvement and development of the range; and the Secretary of the Interior is authorized to continue the study of erosion and flood control and to perform such work as may be necessary amply to protect and rehabilitate the areas subject to the provisions of this subchapter, through such funds as may be made available for that purpose, and any willful violation of the provisions of this subchapter or such rules and regulations thereunder after actual notice thereof shall be punishable by a fine of not more than $500.

# Federal Land Policy and Management Act of 1976, as Codified

### Title 43, Chapter 35, Subchapter I, Section 170. Congressional Declaration of Policy

(A) The Congress declares that it is the policy of the United States that

1. The public lands be retained in Federal ownership, unless as a result of the land use planning procedure provided for in this Act, it is determined that disposal of a particular parcel will serve the national interest;
2. The national interest will be best realized if the public lands and their resources are periodically and

systematically inventoried and their present and future use is projected through a land use planning process coordinated with other Federal and State planning efforts;

3. Public lands not previously designated for any specific use and all existing classifications of public lands that were effected by executive action or statute before October 21, 1976, be reviewed in accordance with the provisions of this Act;

4. The Congress exercise its constitutional authority to withdraw or otherwise designate or dedicate Federal lands for specified purposes and that Congress delineate the extent to which the Executive may withdraw lands without legislative action;

5. In administering public land statutes and exercising discretionary authority granted by them, the Secretary be required to establish comprehensive rules and regulations after considering the views of the general public; and to structure adjudication procedures to assume adequate third party participation, objective administrative review of initial decisions, and expeditious decision making;

6. Judicial review of public land adjudication decisions be provided by law;

7. Goals and objectives be established by law as guidelines for public land use planning, and that management be on the basis of multiple use and sustained yield unless otherwise specified by law;

8. The public lands be managed in a manner that will protect the quality of scientific, scenic, historical, ecological, environmental, air and atmospheric, water resource, and archaeological values; that, where appropriate, will preserve and protect certain public lands in their natural condition; that will provide food and habitat for fish and wildlife and domestic animals; and that will provide for outdoor recreation and human occupancy and use;

9. The United States receive fair market value of the use of the public lands and their resources unless otherwise provided for by statute;

10. Uniform procedures for any disposal of public land, acquisition of non-Federal land for public purposes, and

the exchange of such lands be established by statute, requiring each disposal, acquisition, and exchange to be consistent with the prescribed mission of the department or agency involved, and reserving to Congress review of disposals in excess of a specified acreage;

11. Regulations and plans for the protection of public land areas of critical environmental concern be promptly developed;

12. The public lands be managed in a manner which recognizes the Nation's needs for domestic sources of minerals, food, timber, and fiber from the public lands including implementation of the Mining and Minerals Policy Act of 1970 as it pertains to the public lands; and

13. The Federal Government should, on a basis equitable to both the Federal and local taxpayers, provide for payments to compensate States and local governments for burdens created as a result of the immunity of Federal lands from State and local taxation.

# Reports

When a conflict arises, one of the ways that policy-makers seek resolution is by looking at various types of information that might help inform the debate. The Government Accountability Office (GAO), formerly known as the General Accounting Office, produces reports for Congress, the executive branch, and the public on topics of current political interest. Most of the reports deal with how well (or how poorly) a policy is working, and the studies they conduct provide information that might not otherwise be available. The reports are also valuable because they begin with a neutral statement of the issues at hand, a brief summary of the methodology used, and comments from the affected agencies on a draft of the report before it is released. In September 2005, the GAO produced a report on rangeland management, and more specifically, on the purpose of grazing fees and how much the federal government receives in income. Although the GAO generally provides only its appraisal of program operations, in this case it does so in a way that points out financial considerations that Congress potentially needs to consider.

## Abstract, U.S. General Accounting Office Report, GAO-05–869, *Livestock Grazing: Federal Expenditures and Receipts Vary, Depending on the Agency and the Purpose of the Fee Charged* (September 30, 2005)

Ranchers pay a fee to graze their livestock on federal land. Grazing occurs primarily on federal land located in the Western states managed by ten federal agencies. Generally, the fee is based on animal unit months (AUM)—the amount of forage that a cow and her calf can eat in one month. For most federal land, the fee per AUM is established by a formula. Advocates argue that grazing uses federal land productively and that the grazing fee is fair. Opponents argue that grazing damages public resources and that grazing fees are too low. The GAO was asked to determine (1) the extent of, and purposes for, grazing in fiscal year 2004 on lands that ten federal agencies manage; (2) the amount that federal agencies spent in fiscal year 2004 to manage grazing; (3) total grazing receipts that the ten agencies collected in fiscal year 2004 and amounts disbursed; and (4) fees charged by the ten federal agencies, Western states, and ranchers, and reasons for any differences. In commenting on a draft of this report, the Department of the Interior and the Forest Service neither agreed nor disagreed with the findings. The Forest Service stated that the report accurately described the purpose of the grazing fee. The Army and Air Force and the Department of Energy provided technical comments, which we incorporate as appropriate. The departments of Commerce and of Justice responded that they did not have comments.

The ten federal agencies managed more than 22.6 million AUMs on about 235 million acres of federal lands for grazing and land management in fiscal year 2004. Of that total, the Department of the Interior's Bureau of Land Management (BLM) and the U.S. Department of Agriculture's Forest Service managed more than 98 percent of the lands used for grazing. The agencies manage their grazing programs under different authorities and for different purposes. For BLM lands and Western Forest Service lands, grazing is a major program; the other eight agencies generally use grazing as a tool to achieve their primary land management goals. In fiscal year 2004, federal agencies spent a total of at least $144 million. The ten federal agencies spent at least $135.9 million, with the Forest Service and the BLM accounting for the majority. Other federal agencies have grazing-related activities, such as pest control, and spent at least $8.4 million in fiscal year 2004. The ten federal agencies' grazing fees generated about $21 million in fiscal year 2004—less than one-sixth of the expenditures to manage grazing. Of that amount, the agencies distributed about $5.7 million to states and counties in which grazing occurred, returned about $3.8 million to the Treasury, and deposited at least $11.7 million in separate Treasury accounts to help pay for agency programs, among other things. The amounts that each agency distributed varied, depending on the agencies' differing authorities. Fees charged in 2004 by the ten federal agencies, as well as state land agencies and private ranchers, varied widely. The grazing fee that the BLM and the Forest Service charge, which was $1.43 per AUM in 2004, is established by formula, and is generally much lower than the fees charged by the other federal agencies, states, and private ranchers. The other agencies, states, and ranchers generally established fees to obtain the market value of the forage. The formula used to calculate the BLM and Forest Service grazing fee incorporates ranchers' ability to pay; therefore the current purpose of the fee is not primarily to recover the agencies' expenditures or to capture the fair market value of forage. As a result, the BLM's and the Forest Service's grazing receipts fell short of their expenditures on grazing in fiscal year 2004 by almost $115 million. The BLM and Forest Service fee also decreased by 40 percent from 1980 to 2004, while grazing fees charged by private ranchers increased by 28 percent during the same period. If the purpose of the fee were to recover expenditures, BLM and the Forest Service would have had to charge $7.64 and $12.26 per AUM, respectively; alternatively, if the purpose were to gain a fair market value, the agencies' fees would vary depending on the market. Differences in resources and legal requirements can cause fees to vary; however, the approaches used by other agencies could close the gap in expenditures and receipts or more closely align BLM and Forest Service fees with market prices. The purpose of the grazing fee is, ultimately, for Congress to determine.

# Resolutions

Unlike legislation, resolutions are statements of intent or planned action, and they usually are an expression of sentiments or opinion, rather than direction. The first example, directed to the president of the United States, is a memorial (another term for a resolution) by a professional organization, the American Association for the Advancement of Science. This excerpt deals with the management of timber resources, resulting from an 1887 Department of Agriculture decision to create a division of forestry to collect information on the value and significance of forest cover. The association argues that timber lands have been poorly managed, with insufficient attention to the role of forest cover as part of the watershed. The members approach the problem, they note, from a scientific perspective, with specific recommendations on the need for forest reserves. Also of importance is the recommendation that existing departments and bureaus within the federal government cooperate on the development of a commission to study forest management, and the withdrawal of forest lands until the data is collected and analyzed.

The second resolution also comes from a professional organization; it deals with the problems associated with the sale of diamonds acquired illegally or as a result of civil unrest. The diamond industry, composed of businesses handling rough diamonds, processing, exporting, and importing, has tried to police its own members. Diamond industry members issued a joint resolution on July 19, 2000, as part of their international meeting in Antwerp, Belgium. An excerpt from the resolution illustrates the perspective of the industry in nine separate proposals.

The United Nations and individual states have attempted to resolve problems on a political level, which is often difficult given the uncertainty of power in individual countries. In May 2000, diamond-producing states in southern Africa met in Kimberley, South Africa, to try to develop a long-term proposal that would ensure consumers that diamonds they purchase were not associated with violent conflict. Two years later, after discussions and negotiations that were sometimes bitter, the UN General Assembly adopted a resolution supporting an international certification process for diamonds, the Kimberley Process Certification Scheme (KPCS). The preamble to the agreement identifies the rationale behind the system, which calls for voluntary self-regulation.

Lastly, the Bonn Ministerial Declaration is a statement by international water leaders who met at the International Conference on Freshwater in Bonn, Germany, prior to the 2002 World Summit for Sustainable Development, held in Johannesburg, South Africa. Meetings like this one are examples of international efforts to deal with water scarcity in a formal, international forum. Previous meetings had been held in Dublin, Paris, France, and The Hague, and subsequent efforts to create an international agreement on water and sustainability have been held periodically since then.

## Excerpt, Memorial of the American Association for the Advancement of Science in Behalf of a Proper Forest Policy

*Resolved,* That it is the sense of the American Association for the Advancement of Science that immediate action should be taken looking to the establishment of a proper administration of the remaining timber lands in the hands of the Governments of the United States and Canada, for the purpose of insuring the perpetuity of the forest cover on the western mountain ranges, preserving thereby the dependent favorable hydrological conditions.

*Resolved further,* That a committee of five be, and is hereby, appointed to present this resolution, and to urge the importance thereof to the President and the Congress of the United States, and to the premier and Parliament of Canada and of the provincial governments, and that such committee be instructed to prepare in proper form any data necessary, and to use every honorable means to accomplish the purpose herein set forth, and that the president of this association be hereby appointed chairman of such committee, together with four others whom he shall appoint.

The administration of the timber lands has been unsatisfactory for lack of proper legislation and of provisions sufficient to protect this property against material loss and deterioration. Timber thieving and destruction by fire have been allowed to unnecessarily waste this national property, while the officers in charge were powerless to protect it. The pioneer legislation, which may have been sufficient twenty-five years ago, has long outlived its usefulness and should make way for such administration as will meet the demands of civilized existence in settled communities.

A vast empire, considered useless not long ago, has been found capable of human occupancy and agricultural production if the means for its development, water, can be brought upon it, and the extent to which this land may be utilized depends upon the amounts of water available.

The opinions of our greatest climatologists have been divided as to the influence of forests on precipitation. But evidence, carefully and scientifically scrutinized, is accumulating which tends to show that, under certain conditions at least, such influence may not be improbable. However this may be, overwhelming evidence can be brought to show that a potent influence upon the distribution of available water supplies from rain and snow is exerted by a forest cover, so that a government having to deal with the problem of cultural development of a part of its domain by agriculture, cannot compass the water question without at the same time giving attention and proper regard to the forestry question.

Removal of the mountain forest means invariably disturbance of the natural "run off"; favorable sometimes, unfavorable mostly.

It may be difficult to devise at once such a plan for the administration of these forests, with a view to their continuity, as can be put into practice under the present social and political conditions of that part of our country in which this timbered area is situated; and a special investigation of these conditions and careful adjustment between the present needs of the population for wood material and the future needs of a forest cover for hydrologic purposes appears desirable, although various measures for a forest administration which seem capable of practical application have been proposed.

We, therefore, the undersigned committee, in the conservative and scrutinizing spirit that should characterize the proposition of the scientific body which we represent respectfully recommend:

That a joint committee of the Senate and House of Representatives of the United States be appointed to consider the needs of legislation in behalf of the public timber domain with a view of providing for the appointment of a commission of competent men, salaried and employed for this service alone, for the purpose of investigating the necessity of preserving certain parts of the present public forest area as requisite for the maintenance of favorable water conditions, and to devise a practical plan for the permanent administration of such parts of it as shall appear desirable to be retained under government control.

The committee further recommends that, pending such investigation, all timber lands now in the hands of the United States be withdrawn from sale and provision be made to protect the said lands from theft and ravages by fires and to supply in a rational manner the local needs for wood and lumber until a permanent system of forest administration be had.

It is also suggested that inasmuch as the various Departments and Government bureaus, namely the Department of Agriculture in its Forestry Division, the Department of the Interior in its Land Office and Geological Survey, the Department of War in its Signal Office, the Treasury Department in its Coast and Geodetic Survey, are more or less closely interested in this matter and have collected data useful in the work of such a commission, these Departments should cooperate and act as advisers of said Commission.

## Joint Resolution, World Federation of Diamond Bourses (WFDB) and International Diamond Manufacturers Association (IDMA)

WFDB and IDMA, representing all the principal diamond manufacturing and trading centers of the world, have consistently been aware of and been involved in combating the conflict diamonds problem. Particularly, they point to the numerous resolutions passed by themselves and their members.

We believe that more can and should be done to limit, if not eliminate, this problem entirely. We believe that the solution to the conflict diamonds problem is a moral imperative above all others. However, we do not believe that the solution necessarily entails damage or limitation to the 96+% of the world's diamond trade which is legitimate. On the contrary, we believe that an enlightened and effective approach to the problem can lead to the improvement of the diamond market overall.

It is our understanding that all concerned parties are aware of the positive benefits of diamonds as well as their potential role in providing prosperity, a key ingredient of peace, in countries currently experiencing strife. Over the past year, various solutions have been proposed. We have analyzed these proposals, some of which we have found to be ineffective, others more practical and some impractical. All of the proposals have had elements that we believe are logical and should be incorporated into an effective solution.

As diamond manufacturers and traders primarily responsible for the conversion of rough diamonds into polished and the marketing of those polished diamonds, we are proposing a number of concrete steps to be taken by all parties concerned which we believe will lead to a more effective and immediate resolution of the problem.

While our proposal may be subject in the future to any number of improvements, we believe it is, in the first instance, practically implementable in the short term, and it does not preclude further steps from being taken as and when the means and requirement arise.

Specifically, and most importantly, we are mindful that the next phase of solution must start sooner rather than later and that if this is to be done in a nondestructive manner, the most practically implementable steps must be taken first, in order that the process not be delayed with theoretical concepts and technologies.

1.  We recognize that rough diamonds individually are not sufficiently determinable as to source and origin. However, with the correct system, rough diamond parcels can be monitored with a net.
2.  There is no implementable means of tagging, tracking and identifying finished polished diamonds.
3.  All legitimate diamonds in the rough form can travel within an identifiable net.

Accordingly, we propose:

1.  Each accredited rough diamond importing country, whether a producer, manufacturing or dealing center enacts "redline" legislation. As such, no parcel of rough may be imported unless such parcel of rough has been sealed and registered in a universally standardized manner by an accredited export authority from the exporting country.
2.  Each exporting country, which can either be a producer country or accredited dealing/manufacturing center, will establish accredited export officers or diamond board which will seal parcels of rough diamonds to be exported and registered in an international database. If the country is a producer country, it will be accredited only if it has control mechanisms in place to determine the flow of rough and legitimate ownership of rough presented to the export authority.

3. Polished diamond consuming countries will enact legislation forbidding importation of polished diamonds from any manufacturing/dealing country that does not have "redline" legislation as regards the importation of rough.
4. Each and every country, as part of the diamond net, be they rough exporters, importers, or polished consuming countries enacts legislation bringing criminal penalties on any individual and/or company proven to be knowingly involved in illegal rough diamonds.
5. Each and every diamond organization adopts an ethical code of conduct as regards conflict diamonds, labor practices and good business practices in general, the failure to adhere to which would lead to expulsion from WFDB, IDMA and all other relevant organizations.
6. As a positive measure of compliance, all relevant and interested parties promote adherence to the code of conduct as a positive consumer choice in the marketplace.
7. We enlist the support of banks, insurance, shipping companies and other pertinent providers of goods and services to our industry to expose and cease business relations with any entity that is found knowingly to violate these principles.
8. That there is a continual analysis of relevant technologies and investment by the industry in developing them further for implementation leading to greater compliance.
9. That compliance with the above be monitored and controlled by an international Diamond Council comprised of producers, manufacturers, traders, governments and relevant international organizations. That this process be fully verified and audited.

# Preamble to the Kimberley Process Certification Scheme

PARTICIPANTS,

RECOGNISING that the trade in conflict diamonds is a matter of serious international concern, which can be directly linked to the fueling of armed conflict, the activities of rebel movements aimed at undermining or overthrowing legitimate governments, and the illicit traffic in, and proliferation of, armaments, especially small arms and light weapons;

FURTHER RECOGNISING the devastating impact of conflicts fueled by the trade in conflict diamonds on the peace, safety, and security of people in affected countries and the systematic and gross human rights violations that have been perpetrated in such conflicts;

NOTING the negative impact of such conflicts on regional stability and the obligations placed upon states by the United Nations Charter regarding the maintenance of international peace and security;

BEARING IN MIND that urgent international action is imperative to prevent the problem of conflict diamonds from negatively affecting the trade in legitimate diamonds, which makes a critical contribution to the economies of many of the producing, processing, exporting and importing states, especially of developing states;

RECALLING all of the relevant resolutions of the United Nations Security Council under Chapter VII of the United Nations Charter, including the relevant provisions of Resolutions 1173 (1998), 1295 (2000), 1306 (2000), and 1343 (2001), and determined to contribute to and support the implementation of the measures provided for in these resolutions;

HIGHLIGHTING the United Nations General Assembly Resolution 55/56 (2000) on the role of the trade in conflict diamonds in fueling armed conflict, which called on the international community to give urgent and careful consideration to devising effective and pragmatic measures to address this problem;

FURTHER HIGHLIGHTING the recommendation in United Nations General Assembly Resolution 55/56 that the international community develop detailed proposals for a simple and workable international certification scheme for rough diamonds based primarily on national certification schemes and on internationally agreed minimum standards;

RECALLING that the Kimberley Process, which was established to find a solution to the international problem of conflict diamonds, was inclusive of concerned stakeholders, namely producing, exporting and importing states, the diamond industry, and civil society;

CONVINCED that the opportunity for conflict diamonds to play a role in fueling armed conflict can be seriously reduced by introducing a certification scheme for rough diamonds designed to exclude conflict diamonds from the legitimate trade;

RECALLING that the Kimberley Process considered that an international certification scheme for rough diamonds, based on national laws and practices and meeting internally agreed minimum standards, will be the most effective system by which the problem of conflict diamonds could be addressed;

ACKNOWLEDGING the important initiatives already taken to address this problem, in particular by the governments of Angola, the Democratic Republic of Congo, Guinea, and Sierra Leone and by other key producing, exporting and importing countries, as well as by the diamond industry, in particular by the World Diamond Council, and by civil society;

WELCOMING voluntary self-regulation initiatives announced by the diamond industry and recognising that a system of such voluntary self-regulation contributes to ensuring an effective internal control system of rough diamonds based upon the international certification scheme for rough diamonds;

RECOGNISING that an international certification scheme for rough diamonds will only be credible if all Participants have established internal systems of control designed to eliminate the presence of conflict diamonds in the chain of producing, exporting and importing rough diamonds within their own territories, while taking into account that differences in production methods and trading practices as well as differences in institutional controls thereof may require different approaches to meet minimum standards;

FURTHER RECOGNISING that the international certification scheme for rough diamonds must be consistent with international law governing international trade;

ACKNOWLEDGING that state sovereignty should be fully respected and the principles of equality, mutual benefits and consensus should be adhered to;

RECOMMEND THE FOLLOWING PROVISIONS . . .

# Bonn Ministerial Declaration (2001)

We, ministers with responsibilities for water affairs, environment and development from 46 countries throughout the world, have assembled in Bonn to assess progress in implementing Agenda 21 and to discuss actions required to increase water security and to achieve sustainable management of water sources.

We consider that the World Summit for Sustainable Development, scheduled for August 2002 in Johannesburg, needs to demonstrate renewed commitment to sustainable development and political will to action.

We consider the equitable and sustainable use and the protection of the world's freshwater resources a key challenge facing governments on the road to a safer, more peaceful, equitable and prosperous world. Combating poverty is the main challenge for achieving equitable and sustainable development, and water plays a vital role in relation to human health, livelihood, economic growth as well as sustaining ecosystems. The outcome of the World Summit on Sustainable Development must include decisive action on water issues.

We express our deep concern that at the beginning of the 21st century 1.2 billion people live a life in poverty without access to safe drinking water, and that almost 2.5 billion have no access to proper sanitation. Safe and sufficient water and sanitation are basic human needs. The worldwide struggle to alleviate poverty must bring safe and decent living conditions to those who are deprived of these basic requirements.

We confirm our resolve to reach the International Development Targets agreed by the U.N. Millennium Summit, in particular the target to halve, by the year 2015, the proportion of people living in extreme poverty and to halve the proportion of people who suffer from hunger and are unable to reach or afford safe drinking water. We also confirm our resolve to stop the unsustainable exploitation of water resources by developing water management strategies at regional, national and local levels.

Water is needed in all aspects of life. For sustainable development, it is necessary to take into account water's social, environmental, and economic dimensions and all of its varied uses. Water management therefore requires an integrated approach.

We emphasize that ten years after the U.N. Conference on Environment and Development and the Dublin Conference, and several years after the global water conferences in Paris and The Hague, there is still a need for greater commitment to implement commonly agreed principles on water resource management. Pressures on the world's scarce freshwater resources and aquatic systems have increased. Water pollution and unsustainable patterns of water consumption are among the causes. Water use efficiency needs to improve.

We agree that governments, the international community, the private sector, the nongovernmental organizations and all other stakeholders need to base their actions on the following:

Governance: The primary responsibility for ensuring the sustainable and equitable management of water resources rests with the governments. Each country should have in place applicable arrangements for the governance of water affairs at all levels and, where appropriate, accelerate water sector reforms. We urge the private sector to join with government and civil society to contribute to bringing water and sanitation services to the unserved and to strengthen investment and management capabilities. Privately managed service delivery should not imply private ownership of water resources. Service providers should be subject to effective regulation and monitoring. We encourage riparian states to cooperate on matters related to international watercourses.

Funding Gap: There is an enormous gap in funding investments in water infrastructure, maintenance, training and capacity building, research, and data generation. It is urgent to close this gap using existing resources more efficiently and with additional financial resources from all sources: public investment budgets, capital markets, and community based finance, user and polluter charges; as well as increased international development financing from public and private sources, particularly for developing countries to reflect the acute needs in the water sector. The lack of financial resources for water infrastructure investment, operations and maintenance is particularly hurting the poor in Least Developed Countries and in other countries with people living in extreme poverty. Critical actions for closing the financial gap are poverty alleviation and the improvement of opportunities for trade and income generation for developing countries. Resources also need to be made available to assist developing countries to mitigate the effects of natural disasters and to assist in adapting to the impacts of climate change. Water development programs, to be successful, should be based on a good understanding of the negative impact of desertification causes to people living in affected areas.

Role of the International Community: We call on the international community to strengthen its commitment and its efforts to enable developing countries to manage water sustainably and to ensure an equitable sharing of benefits from water resources. We call upon the Secretary General of the United Nations to strengthen the coordination and coherence of activities within the U.N. system on water issues in an inclusive manner. We recall the agreed U.N. target for official development assistance to 0.7% of GDP. Developed countries which have not yet reached the target should exert their best efforts to do so.

Capacity Building and Technology Transfer: We recognize that capacity building and innovative technologies are needed to efficiently utilize water, control pollution and develop new and alternative water sources in water stressed countries. We will support capacity building programs and information exchange to ensure the effective use of human, financial, and technical resources for water management. We will facilitate technology transfer initiatives to enable technologically less developed countries to acquire capacity to manage water with the best available knowledge and equipment. We need improved assessments of state and trends in the world water situation.

Gender: Water resources management should be based on a participatory approach. Both men and women should be involved equally in managing the sustainable use of water resources and sharing of benefits. The role of women in water related areas needs to be strengthened and their participation broadened.

Next Steps: We urge the World Summit on Sustainable Development to take account of the outcome of this International Conference on Freshwater. We expect that the International Year of Freshwater in 2003, and the 3rd World Water Forum in Japan will be a good opportunity to further discuss the role and actions for all players in international society on the issues of sustainable development of freshwater.

# Facts and Data

## U.S. Natural Resource Agencies

In the nation's infancy, the government was concerned about natural resources and how best to use them to help the country grow, become settled, and later, industrialized. The executive branch has grown considerably in its oversight of natural resources, with dozens of agencies and bureaus assigned the task of dealing with problems and solutions. Table 6.1 shows how fragmented natural resource management has become in the twenty-first century.

**TABLE 6.1**
**U.S. Natural Resource Agencies (Partial listing)**

| | |
|---|---|
| Bureau of Indian Affairs | Nuclear Regulatory Commission |
| Bureau of Land Management | Office of Conservation and Renewable Energy |
| Bureau of Reclamation | Office of Environmental Justice |
| Council on Environmental Quality | Office of Environmental Policy |
| Council on Sustainable Development | Office of Information and Regulatory Affairs |
| Environmental Protection Agency | Office of Management and Budget |
| Federal Energy Regulatory Commission | Office of Surface Mining Reclamation and Enforcement |
| Government Accountability Office | U.S. Army Corps of Engineers |
| Materials Transportation Bureau | U.S. Department of Agriculture |
| Minerals Management Service | U.S. Department of Commerce |
| Mine Safety and Health Administration | U.S. Department of Energy |
| National Bureau of Standards | U.S. Department of the Interior |
| National Center for Environmental Economics | U.S. Department of State |
| National Institute for Occupational Safety and Health | U.S. Geological Survey |
| National Oceanic and Atmospheric Administration | U.S. Fish and Wildlife Service |
| National Park Service | U.S. Forest Service |

# National Landscape Conservation System

Some of the reasons why many of the conflicts over protected areas are centered in the West are the number of sites, the acreage, and the mileage included in the National Landscape Conservation System (NLCS). As Table 6.2 indicates, the majority of units are concentrated in just twelve states, with California and Utah ranked the highest in terms of acreage. Of particular interest are the wilderness areas and wilderness study areas, which receive the most protected status. Three wilderness boundaries cross state lines; national scenic and historic trails also cross state lines. Some of the acreage figures represent significant overlaps, such as Wild and Scenic Rivers located within Wilderness Areas. The Bureau of Land Management (BLM) has responsibility for a vast amount of landscape, making it one of the more powerful agencies in the federal government when it comes to range and protected area management.

**TABLE 6.2**
**NLCS Sites, By State (As of July 2005)**

| State | Number | BLM Acres |
|---|---|---|
| **National Conservation Areas** | | |
| Alaska | 1 | 1,208,624 |
| Arizona | 3 | 112,542 |
| California | 2 | 10,728,368 |
| Colorado | 2 | 185,144 |
| Idaho | 1 | 484,034 |
| Nevada | 3 | 1,043,422 |
| New Mexico | 1 | 339,100 |
| Total | 13 | 14,101,234 |
| **National Monuments** | | |
| Arizona | 5 | 1,775,007 |
| California | 3 | 291,390 |
| Colorado | 1 | 163,892 |
| Idaho | 1 | 273,847 |
| Montana | 2 | 375,027 |
| New Mexico | 1 | 4,124 |
| Oregon/Washington | 1 | 52,947 |
| Utah | 1 | 1,870,800 |
| Total | 15 | 4,807,034 |
| **Cooperative Management and Protection Areas** | | |
| Oregon/Washington | 1 | 428,156 |
| **National Recreation Areas** | | |
| Alaska | 1 | 998,702 |
| **Outstanding Natural Areas** | | |
| Oregon/Washington | 1 | 100 |
| **Forest Reserves** | | |
| California | 1 | 7,472 |
| **Wilderness Areas** | | |
| Arizona | 47 | 1,396,586 |
| California | 76 | 3,577,796 |
| Colorado | 4 | 139,524 |
| Idaho | 1 | 802 |
| Montana | 1 | 6,000 |
| Nevada | 38 | 1,758,606 |
| New Mexico | 3 | 139,281 |

*(continues)*

**TABLE 6.2** *(cont.)*
**NLCS Sites, By State (As of July 2005)**

| State | Number | BLM Acres |
|---|---|---|
| **Wilderness Areas (continued)** | | |
| Oregon/Washington | 5 | 193,863 |
| Utah | 3 | 27,720 |
| Total | 178 | 7,240,178 |
| **Wilderness Study Areas** | | |
| Alaska | 1 | 784,238 |
| Arizona | 2 | 63,930 |
| California | 77 | 974,769 |
| Colorado | 54 | 621,737 |
| Idaho | 66 | 1,341,709 |
| Montana | 40 | 450,823 |
| Nevada | 71 | 2,877,917 |
| New Mexico | 60 | 970,532 |
| Oregon/Washington | 98 | 2,343,279 |
| Utah | 99 | 3,255,490 |
| Wyoming | 42 | 575,841 |
| Total | 610 | 14,260,265 |
| **National Historic Trails** | | |
| Alaska | 1 | 418 miles |
| Arizona | 2 | 89 |
| California | 3 | 423 |
| Colorado | 1 | 85 |
| Idaho | 4 | 439 |
| Montana | 2 | 347 |
| Nevada | 3 | 1,065 |
| New Mexico | 2 | 156 |
| Oregon/Washington | 2 | 24 |
| Utah | 3 | 569 |
| Wyoming | 5 | 1,262 |
| Total | 10* | 4,877 |
| **National Scenic Trails** | | |
| California | 1 | 189 miles |
| Colorado | 1 | 1 |
| Idaho | 1 | 13 |
| Montana | 1 | 11 |

*(continues)*

**TABLE 6.2** *(cont.)*
**NLCS Sites, By State (As of July 2005)**

| State | Number | BLM Acres |
|---|---|---|
| **National Scenic Trails (continues)** | | |
| New Mexico | 1 | 172 |
| Oregon/Washington | 1 | 42 |
| Wyoming | 1 | 180 |
| Total | 2* | 608 miles |
| **Wild and Scenic Rivers** | | |
| Alaska | 6 | 952 miles |
| California | 6 | 78 |
| Montana | 1 | 149 |
| New Mexico | 2 | 71 |
| Oregon/Washington | 23 | 802 |
| Total | 38 | 2,052 |

*Source:* U.S. Bureau of Land Management, Office of Public Affairs at www.blm.gov/nlcs (accessed February 14, 2006).

*National Historic and Scenic Trails cross state lines. There are a total of 10 Historic and 2 Scenic Trails in the National Trails System administered by the Bureau of Land Management.

# National Park Visitation

It has been said that America's national parks are "being loved to death" because of their popularity, and that is certainly the case with specific sites. Increased visitation has put pressures on the system in terms of infrastructure, congestion, air quality, noise, and staff resources. A look at the statistics (see Table 6.3) compiled by the National Park Service (NPS) shows that the total number of visits between 1980 and 2005 has grown considerably, but there have been some years when recreational and nonrecreational visits have actually declined in number. The period from 1999 to 2005 shows declining visitation, which has been attributed to reluctance by visitors to travel after the terrorist attacks on the World Trade Center and the Pentagon on September 11, 2001.

# Ten Most Endangered Birds in America

The National Audubon Society has worked since the late 1800s to protect birds and their habitats. As one of the nation's oldest environmental organizations, the society is known for its efforts to protect endangered species both in the United States and around

**TABLE 6.3**
**National Park System Visitation, By Year (1980–2005)**

| Year | Total Visits | Year | Total Visits |
|------|-------------|------|-------------|
| 1980 | 300,324,082 | 1993 | 387,707,068 |
| 1981 | 329,663,300 | 1994 | 380,156,046 |
| 1982 | 334,448,015 | 1995 | 387,803,913 |
| 1983 | 335,646,331 | 1996 | 399,826,439 |
| 1984 | 332,671,283 | 1997 | 418,161,349 |
| 1985 | 346,190,110 | 1998 | 435,660,070 |
| 1986 | 364,551,499 | 1999 | 436,295,878 |
| 1987 | 372,951,134 | 2000 | 429,853,123 |
| 1988 | 367,989,781 | 2001 | 424,302,551 |
| 1989 | 351,911,180 | 2002 | 421,279,444 |
| 1990 | 335,162,737 | 2003 | 413,865,726 |
| 1991 | 355,868,408 | 2004 | 427,706,748 |
| 1992 | 360,351,945 | 2005 | 423,387,198 |

*Source:* National Park Service at www2@nature.nps.gov/NPstats (accessed November 15, 2005).

the world. One of its successful strategies has been to identify birds that are in need of special protection and to focus public and media attention on the factors leading toward extinction. In 2006, attempts within Congress to weaken the provisions of the Endangered Species Act led to the publication of a National Audubon Society Report that names the country's most endangered birds (see Table 6.4). The report cites pressure from development, invasive species, and global warming as the primary threats to these species, moving them closer to extinction.

## U.S. Most Endangered Wildlife Refuges

One of the nation's most influential environmental organizations, Defenders of Wildlife, works to protect both wildlife and their habitats, including the nation's system of National Wildlife Refuges. These 545 protected areas, initially established by President Theodore Roosevelt, are considered endangered because of a lack of adequate resources, proposed encroachment by development, and plans for oil and natural gas drilling. Two hundred

**TABLE 6.4**
**Ten Most Endangered Birds in America (2006)**

| | |
|---|---|
| Ivory-billed Woodpecker | California Condor |
| Whooping Crane | Gunnison Sage-Grouse |
| Kirtland's Warbler | Piping Plover |
| Florida Scrub-Jay | Ashy Storm-Petrel |
| Golden-Cheeked Warbler | Kittlitz's Murrelet |

*Source:* "National Audubon Society Report Lists America's 10 Most Endangered Species," news release (March 27, 2006) at www.audubon.org/news/press (accessed April 9, 2006).

of the refuges have no staff at all, and others are operating on a fraction of their previous budgets.

In its October 2005 report, *Refuges at Risk: America's Ten Most Endangered National Wildlife Refuges 2005,* the group identified the ways in which complex political and environmental issues are changing both the operations and health of these protected areas. While not nearly as well known as the national parks, wildlife refuges protect areas that otherwise might not be appropriate for designation under existing policies. The 2005 list, in alphabetical order, identifies the ten most endangered refuges, representing ten states in regions throughout the United States (see Table 6.5).

## U.S. Gold Prices

Conflicts over gold are not new, but they have escalated since the 1970s. Gold has been valuable since antiquity, but a uniform price for gold was first established by Sir Isaac Newton, as master of the United Kingdom Mint, in 1717. In U.S. dollars, the average price of gold changed by only a few dollars from 1792 to 1930. Initially, the price was $19.75 per troy ounce in 1792; it had risen to $18.96 by 1892, and to $20.66 in 1922. Real fluctuations rarely occurred, except in 1931, when the price dropped to $17.06; throughout the rest of the 1930s and into the 1960s, the price of gold remained at about $35. A two-tiered pricing system was created in 1968, and after that, the market price for gold has been free to fluctuate (see Table 6.6).

**TABLE 6.5**
**Ten Most Endangered National Wildlife Refuges (2005)**

**Arctic National Wildlife Refuge, AK**
Drilling proponents have resorted to years of legislative maneuvering to allow for drilling, even though there is strong public opposition because of threats to wildlife and the pristine wilderness..

**Browns Park National Wildlife Refuge, CO**
Livestock grazing has degraded the riparian area along the Green River, home to thousands of migratory birds, elk, and deer.

**Buenos Aires National Wildlife Refuge, AZ**
Thousands of migrant border crossings and government enforcement agent activities have severely damaged the fragile desert and its wildlife, including the endangered masked bobwhite quail.

**Florida Panther National Wildlife Refuge, FL**
One of the world's most endangered species is at risk as roads and other developments encroach upon the habitat of the Florida panther.

**McFaddin National Wildlife Refuge, TX**
The expansion of oil and gas drilling near Texas's largest freshwater marsh threatens the wintering and resting area for migrating birds.

**Mingo National Wildlife Refuge, MO**
A proposed power plant scheduled to be built upwind of the refuge will send pollutants through a pristine bottomland hardwood forest.

**Moapa National Wildlife Refuge, NV**
As Las Vegas and its suburbs have grown, groundwater pumping is reducing water sources, including desert springs and the species that depend upon them.

**Oyster Bay National Wildlife Refuge, NY**
Sewage discharges from motorized pleasure boats and storm water runoff are damaging this estuary, located near the home of Theodore Roosevelt, the father of the nation's wildlife refuge system.

**Pocosin Lakes National Wildlife Refuge, NC**
A proposed landing field will endanger the home of more than 100,000 birds, while also endangering military aircraft and pilots.

**Sonny Bono Salton Sea National Wildlife Refuge, CA**
Unless the government embarks immediately on a habitat restoration plan, a proposed water transfer will destroy bird habitat along the important Pacific Flyway.

*Source:* Defenders of Wildlife, "Refuges at Risk: America's Ten Most Endangered National Wildlife Refuges 2005" (October 2005) at www.defenders.org/refugesatrisk (accessed November 19, 2005).

**TABLE 6.6**
**Historic Average Gold Prices 1968–2005 (In U.S. dollars per troy ounce)**

| Year | US$ | Year | US$ |
|------|-----|------|-----|
| 1968 | 39 | 1987 | 447 |
| 1969 | 41 | 1988 | 437 |
| 1970 | 36 | 1989 | 381 |
| 1971 | 40 | 1990 | 384 |
| 1972 | 58 | 1991 | 362 |
| 1973 | 97 | 1992 | 344 |
| 1974 | 154 | 1993 | 360 |
| 1975 | 161 | 1994 | 384 |
| 1976 | 125 | 1995 | 384 |
| 1977 | 148 | 1996 | 388 |
| 1978 | 193 | 1997 | 331 |
| 1979 | 306 | 1998 | 294 |
| 1980 | 615 | 1999 | 279 |
| 1981 | 460 | 2000 | 279 |
| 1982 | 376 | 2001 | 271 |
| 1983 | 424 | 2002 | 310 |
| 1984 | 361 | 2003 | 363 |
| 1985 | 317 | 2004 | 410 |
| 1986 | 368 | 2005 | 445 |

**TABLE 6.7**
**Extent of Worldwide Forest and Wooded Land By Region (2005)**

| Region | 1000 hectares | % of land area |
|---|---|---|
| Eastern and Southern Asia | 226,534 | 27.8 |
| Northern Africa | 131,048 | 8.6 |
| Western and Central Africa | 277,829 | 44.1 |
| Total: Africa | 635,411 | 21.4 |
| | | |
| East Asia | 244,862 | 21.3 |
| South and Southeast Asia | 283,127 | 33.4 |
| Western and Central Asia | 43,626 | 4.0 |
| Total: Asia | 571,615 | 18.5 |
| | | |
| Total: Europe | 1,001,394 | 44.3 |
| Total: Caribbean | 5,974 | 26.1 |
| Total: Central America | 22,411 | 43.9 |
| Total: North America | 677,454 | 32.7 |
| Total: Oceania | 206,252 | 24.3 |
| Total: Latin America | 831,540 | 47.7 |
| Total: World | 3,952,052 | 30.3 |

*Source:* UN Food and Agriculture Organization, at www.fao.org/forestry (accessed February 1, 2006).

# Extent of Worldwide Forest and Wooded Land

The salience, or importance, of a problem depends in large part on how much a segment of the population is affected. Exposure to a natural resource conflict, for example, will increase the interest of the public and political officials because there is more knowledge, media coverage, and impact on those involved. In the case of controversies over forests, statistics alone do not tell the complete story. As Table 6.7 shows, there are more hectares of forest and wooded land in Europe, which has the second largest percentage of total land area as forest. However, most of that land is second-, third-, or even fourth-growth trees, rather than ancient forest. Conflicts over logging are modest in comparison to regions such as North America, where controversy is unrelenting. Parts of the Pacific Northwest and Canada are home to old growth trees, the most valuable from both a resource perspective and an ecological one. Yet less than a third of the land mass is forested. In regions such as Latin America, tropical forests make up almost half of the land area, yet efforts to stop logging, particularly in the Amazon, are relatively recent and often at the urging of non-native groups. Numbers can provide background, but not always the political context of an issue.

## Gas and Oil Supplies

Fossil fuels, especially natural gas and petroleum products, drive international development, whether the fuel is used to provide power and electricity for industry, to light our homes at night, to power our cars, trucks, and other vehicles, or to be made into other products we rely upon as consumers. Supplies of gas and oil are not evenly distributed on the planet, nor is the demand for them. Some countries, such as the United States, have a huge demand that is not met by domestic sources, requiring massive imports (see Table 6.8). Some developing nations have supplies that exceed their current demands, and so they export huge amounts to others. The top sources of U.S. crude oil have varied slightly from month to month but usually include five countries: Mexico, Canada, Saudi Arabia, Nigeria, and Venezuela. The top five exporting countries account for 75 percent of U.S. crude oil imports, with similar figures for petroleum (see Table 6.9).

One of the major concerns raised by analysts is the projected increase in energy demand at a time when resources are dwindling, especially for some fossil fuels. What concerns many policymakers is the fact that several countries that regularly provide the United States with crude oil and natural gas are also areas with a high level of political instability. In recent years, conflicts in the Middle East (especially Iran, Saudi Arabia, and the United Arab Emirates) have led to jumps in prices that have been passed along to consumers in the United States and Japan. Even countries such as Venezuela, which produces about 4 percent of the world's total, have undergone upheaval that has threatened supplies for the United States.

**TABLE 6.8**
**Top Fifteen Crude Oil and Petroleum Exporting Countries to U.S.**
**(In thousands of barrels per day, February 2006)**

| Exporting Country | Crude Oil Imports | Rank | Petroleum Imports | Rank |
| --- | --- | --- | --- | --- |
| Mexico | 1,774 | 1 | 1,878 | 2 |
| Canada | 1,700 | 2 | 2,249 | 1 |
| Saudi Arabia | 1,418 | 3 | 1,451 | 4 |
| Nigeria | 1,342 | 4 | 1,377 | 5 |
| Venezuela | 1,175 | 5 | 1,472 | 3 |
| Angola | 464 | 6 | 478 | 6 |
| Iraq | 444 | 7 | 444 | 8 |
| Ecuador | 222 | 8 | 234 | 11 |
| Brazil | 164 | 9 | 200 | 13 |
| Algeria | 163 | 10 | 446 | 7 |
| Kuwait | 152 | 11 | 158 | 15 |
| Colombia | 108 | 12 | n/a | n/a |
| United Kingdom | 82 | 13 | 205 | 12 |
| Chad | 77 | 14 | n/a | n/a |
| Equatorial Guinea | 73 | 15 | n/a | n/a |
| Virgin Islands | n/a | n/a | 318 | 9 |
| Russia | n/a | n/a | 304 | 10 |
| Norway | n/a | n/a | 199 | 14 |

*Source:* Figures derived from Energy Information Agency, at www.eia.doe.gov (accessed April 15, 2006).

**TABLE 6.9**
**Top Ten Producers of Crude Oil, Petroleum Products, and Natural Gas (2002)**
**(In million tons, as percentage of world total)**

| Country | oil | % of total | petroleum | % of total | gas | % of total |
|---|---|---|---|---|---|---|
| Saudi Arabia | 470 | 12.7 | n/a | n/a | 60,262 | 2.2 |
| Russia | 419 | 11.3 | 184 | 5.2 | 608,332 | 22.4 |
| United States | 348 | 9.4 | 816 | 23.2 | 541,779 | 19.9 |
| Iran | 194 | 5.2 | n/a | n/a | 77,923 | 2.9 |
| Mexico | 189 | 5.1 | n/a | n/a | n/a | n/a |
| PR China | 165 | 4.4 | 207 | 5.9 | n/a | n/a |
| Norway | 151 | 4.1 | n/a | n/a | 76,832 | 2.8 |
| Venezuela | 149 | 4.0 | n/a | n/a | n/a | n/a |
| Canada | 138 | 3.7 | 99 | 2.8 | 182,205 | 6.7 |
| UAE | 120 | 3.2 | n/a | n/a | n/a | n/a |
| Japan | n/a | n/a | 203 | 5.8 | n/a | n/a |
| Germany | n/a | n/a | 114 | 3.2 | n/a | n/a |
| India | n/a | n/a | 112 | 3.2 | n/a | n/a |
| Korea | n/a | n/a | 111 | 3.2 | n/a | n/a |
| Italy | n/a | n/a | 96 | 2.7 | n/a | n/a |
| Brazil | n/a | n/a | 85 | 2.4 | n/a | n/a |
| UK | n/a | n/a | n/a | n/a | 108,438 | 4.0 |
| Algeria | n/a | n/a | n/a | n/a | 86,553 | 3.2 |
| Indonesia | n/a | n/a | n/a | n/a | 79,832 | 2.9 |
| Netherlands | n/a | n/a | n/a | n/a | 73,128 | 2.7 |

*Source:* International Energy Agency, *Key World Energy Statistics* (2004), at www.iea.org (accessed April 15, 2006).

**TABLE 6.10**
**Projected Trends in U.S. Energy Demand, 2004–2030**

| Trend | Projection |
|---|---|
| Average Energy Use, Per Person | Increasing faster than in recent years |
| Total Primary Energy Consumption | Increasing by 1.1 percent per year |
| Delivered Energy Consumption, by Fuel | Petroleum and electricity lead growth in consumption |
| U.S. Primary Energy Use | Increase by one-third, led by commercial sector consumption |
| Residential Energy Use, Per Person | Increase in electricity use, with slight declines in use of natural gas and petroleum |
| Household Energy Use | Energy for space heating decreases by 20 percent; increases slightly for space cooling and lighting; decreases slightly for water heating and refrigeration; fastest increase is for electronics and small appliances |
| Energy Efficiency | Increases in efficiency highest in Texas, California, Arizona, and Nevada, accounting for two-thirds of energy efficient home completions |
| Commercial Energy Use | Increases by 1.6 percent per year |
| Commercial Energy Demand | Cooling, water heating, and lighting; increases for use of office equipment, telecommunications equipment, medical imaging equipment, and other |
| Decreases for space heating, space | miscellaneous uses |
| Commercial Energy Use per Square Foot | Increases by 6 percent |

*Source:* Derived from Energy Information Agency, *Annual Energy Outlook* 2006.

# Energy Use and Trends

From time to time, the Worldwatch Institute uses statistics from various sources to try to put various environmental indicators into perspective. It is often difficult for scientists and researchers to put their data into formats that will provide the public with the information that they and policy-makers need, unless they can see those figures in relation to one another. Tables 6.10 and 6.11 provide a context for understanding energy trends and why they are important.

**TABLE 6.11**
**Energy: Matters of Scale**

| | |
|---|---|
| World energy production from oil, 2003 | 148 quadrillion Btu* |
| Energy production from "new" renewable sources (excludes large hydroelectric plants) | 6 quadrillion Btu |
| Energy production from all renewables (includes large hydro) | 33 quadrillion Btu |
| World annual growth in wind generating capacity, 2000–2004 | +28 percent |
| Annual average growth in solar photovoltaic generating capacity | +32 percent |
| Annual average growth in biofuels (ethanol, biodiesel) production | +18 percent |
| Annual average growth in oil production | +1.6 percent |
| Total increase in oil production (1970–2003) | +52 percent |
| Total increase in renewable energy production (excluding large hydro) | +269 percent |
| Annual government subsidies for renewables (European Union and U.S.) | $10 billion |
| Annual subsidies for fossil energy (global) | $150–250 billion |
| Estimated number of jobs in new renewable manufacturing, operations, and maintenance, 2004 | 1.7 million |
| Jobs in oil and natural gas extraction (U.S. only), 2002 | 123,000 |

*Source:* Worldwatch Institute, www.worldwatch.org, *World Watch,* Jan./Feb. 2006: 32.

* British thermal units

## Water Use

Most people think of water as an abundant natural resource that is always available in a never-ending stream of supply. That view is less common in developing and arid countries, where scarcity is a very real problem, both for sanitation and drinking. In the United States, organizations such as the American Water Works Association Research Foundation have tried to put water in perspective by explaining how residential water use varies (see Table 6.12) and using models to forecast rates for future water demand.

**TABLE 6.12**
**Residential End Uses of Water (U.S.)**

---

Households use approximately 146,000 gallons of water annually.

About 42 percent of water is used indoors, and 58 percent is used outdoors.

In households not utilizing water-efficient fixtures, toilets used the most water on a daily basis, averaging 20.1 gallons per person per day.

The largest residential users of water are clothes washers (15 gallons per person per day).

Taking a shower uses approximately 10 gallons of water.

The most water used is between the hours of 5:00 A.M. and 11:00 A.M.

The least water used is between the hours of 11:00 P.M. and 5:00 A.M.

The presence of teenagers tends to increase a household's water usage; the presence of adults working full-time decreases water usage.

---

*Source:* Reprinted from American Water Works Association website, www.awwa.org, by permission. Copyright 2006, American Water Works Association.

# 7

# Directory of Organizations, Associations, and Agencies

**N**atural resources issues and conflicts cross international boundaries, cultures, and politics, as this directory illustrates. Some of the nongovernmental organizations focus on specific resources, such as timber or precious metals, while others have a broad range of environmental problems on their agenda. Included are trade associations and groups that represent industry interests, along with information clearinghouses that tend to be more neutral and objective. There are a growing number of local and regional grassroots groups that deal with specific problems, and only a handful are included here, to serve as examples. Professional associations represent individuals with common educational credentials, such as experts in range management, forestry, and hydrology. The last two sections provide contact and background information for the major U.S. resource agencies at the federal level, followed by the UN programs.

Because names alone do not always represent what an organization does, there is a brief synopsis of the group's history, mission, or membership along with contact information. A brief word of caution: The Internet is a useful tool for research, but it does have limitations. As quickly as a website can be created, it can disappear, sometimes taking an organization with it. Other times, groups change their names, or are subsumed under another related interest. The information provided in this chapter is current as of publication, but it may not be accurate when attempts are made to contact an organization.

Many of the nongovernmental organizations and research institutes dealing with natural resources in countries outside the United States are new, and they often face budget shortfalls that limit their ability to respond to inquiries. For that reason, the contact information includes a mailing address and, where available, a website. Telephone contact information for international organizations is not included because of differences in access codes.

**Adirondack North Country Association**
28 St. Bernard Street
Saranac Lake, NY 12983
(518) 891–6200
www.adirondack.org
Contact: Executive Director

Founded in 1954, the organization works for the development of the Adirondack region, opposing the creation of an Adirondack National Park because it would restrict the industrial and business potential of the area.

**African Diamond Council**
2800 Post Oak Boulevard
Houston, TX 77056
(832) 623–5923
www.africandiamondcouncil.org
Contact: Director

In an attempt to remove the negative stigma of the African diamond-producing countries (Democratic Republic of Congo, Botswana, South Africa, Angola, Namibia, Ghana, Central African Republic, Guinea, Sierra Leone, and Zimbabwe), this council was formed in 2000. It represents the diamond trade industry and addresses issues such as environmental degradation, uncontrolled mining, diamond labeling, and political turmoil on the African continent related to diamonds.

**Alaska Rainforest Campaign**
122 C Street, NW
Washington, DC 20001
(202) 544–0475
www.akrain.org
Contact: Manager

The Alaska Rainforest Campaign (ARC) is a coalition of national and Alaskan conservation groups working to protect the remaining wildlands of the Tongass and Chugach national forests from clear-cutting and other harmful development.

**Aldo Leopold Foundation**
PO Box 77
Baraboo, WI 53913
(608) 355–0279
www.aldoleopold.org
Contact: Director

The author of the influential book *A Sand County Almanac*, Leopold was among the first to conceive of a land ethic of ecological awareness and stewardship of natural resources. After his death, his children established the foundation in 1982 to foster an ethical relationship between the people and the land. The organization manages the original Leopold Farm and serves as a clearinghouse for information on Leopold and his legacy of ideas.

**Alliance for Energy and Economic Growth**
PO Box 1200
Washington, DC 20013–1200
(202) 463–3130
www.yourenergyfuture.org/index.htm
Contact: Executive Director

With more than 1,200 members representing companies that develop, deliver, or consume energy, this organization's mission relates to maintaining energy security, renewing and expanding the energy infrastructure, and ensuring appropriate consideration of the impact of regulatory policies on energy companies.

**Alliance to Save Energy**
1200 Eighteenth Street, NW, Suite 900
Washington, DC 20036
(202) 857–0666
www.ase.org
Contact: President

The ASE was founded in 1977 as a nonprofit coalition of business, governmental, environmental, and consumer leaders focused on the role of energy efficiency as a cost-effective strategy. The alliance has an energy-efficient headquarters building with advanced technologies to make it 50 percent more efficient than the typical office suite.

**Amazon Conservation Association**
1731 Connecticut Avenue, NW
Washington, DC 20009
(202) 234–2356
www.amazonconservation.org
Contact: Executive Director

Since 1999, the Amazon Conservation Association has implemented field-based scientific and socioeconomic programs to integrate research with natural resources and conservation. The group also collaborates with forest users and communities to improve their resource management practices, encouraging and facilitating sustainable development.

**Amazon Conservation Team**
4211 N. Fairfax Drive
Arlington, VA 22203
(703) 522–4684
www.amazonteam.org
Contact: Executive Director

By working in partnership with indigenous people, ACT's mission is to conserve biodiversity, health, and culture in the Amazon and other ecosystems of tropical and subtropical America. The organization also supports the creation of nature reserves to ensure the protection of critical habitat areas.

**American Coal Foundation**
101 Constitution Avenue, NW, Suite 525E
Washington, DC 20001–2133
(202) 463–9785
www.teachcoal.org

Supported by the coal industry, the foundation is a nonprofit organization providing educational materials on how coal can be used in an environmentally sensitive manner as a part of the nation's fuel sources.

**American Conservation Association**
1200 New York Avenue, NW, Suite 400
Washington, DC 20005
(202) 289–2431
(no website)
Contact: Executive Director

This is an educational and scientific research organization whose goal is to advance the cause of conservation of natural resources and to ensure that these resources are available for public use.

**American Forage and Grassland Council**
PO Box 94
Georgetown, TX 78627
(800) 944–2342
www.afgc.org
Contact: Director

The council supports the idea that livestock can be raised economically and in an environmentally sustainable manner through stewardship and sound agricultural practices.

**American Forest Foundation**
1111 Nineteenth Street, NW, Suite 780
Washington, DC 20036
(202) 463–2462
www.affoundation.org
Contact: Staff

Chartered in 1982, AFF is a nonprofit organization whose purpose is to encourage long-term stewardship of natural resources. It coordinates environmental education and a national habitat conservation program, supported by grants from foundations, government agencies, and corporations.

**American Forests**
PO Box 2000
Washington, DC 20013
(202) 955–4500
www.americanforests.org
Contact: Director

Forest restoration and urban forestry are two of the major projects undertaken by the group, which also advocates positions on issues relating to national forests, community forest activities, and ecosystems.

**American Gas Foundation**
400 N. Capitol Street, NW, Suite 400
Washington, DC 20001
(202) 824–7270
www.gasfoundation.org
Contact: Chairman of the Board

As the voice of the gas industry, this trade association strives to be an independent source of information, research, and programs on energy and environmental issues, with a particular emphasis on natural gas. Founded in 1989, the group conducts executive-level forums and events for its 195 local energy utility companies serving more than 56 million homes, businesses, and industries.

**American Grassfed Association**
PO Box 400
Kiowa, CO 80117
(877) 774–7277
www.americangrassfed.org
Contact: President

Through research, outreach, communication, and marketing, the association promotes the use of grass as a source of nutrition for livestock, identifying the benefits of grassland for animals, consumers, and the environment.

**American Hiking Society**
1422 Fenwick Lane
Silver Spring, MD 20910
(301) 565–6704
www.americanhiking.org
Contact: Staff

The AHS represents the interests of thousands of trail users, and cooperates with federal, state, and local agencies and organizations to maintain trails throughout the nation. Much of the work is done by volunteers who repair trails and pathways for AHS members and for other users. AHS lobbies to protect existing trail networks and to expand trails on public lands.

**American Land Conservancy**
250 Montgomery Street, Suite 210
San Francisco, CA 94104
(415) 912–3660
www.alcnet.org
Contact: President

As a private, nonprofit organization, the conservancy supports the development of partnerships to link land trusts, public lands agencies, and private land owners to protect natural resources and provide strategies for effective conservation solutions.

**American Land Rights Association**
PO Box 400
Battle Ground, WA 98604
(360) 687–3087
www.landrights.org

Founded in 1978 as a grassroots organization comprised of national park inholders, the group now works to maintain access to public lands, protect private property rights, and advocate the wise use of resources in the United States.

**American Lands Alliance**
726 Seventh Street, SE
Washington, DC 20003
(202) 547–9400
www.americanlands.org
Contact: Executive Director

American Lands' programs are designed to protect and restore functioning ecosystems and address the underlying threats to them. Its programs rely on science-based solutions for healing native forests, grasslands, and wetlands by providing a clearinghouse on information and facilitating partnerships among groups.

**American Water Resources Association**
4 West Federal Street
Middleburg, VA 20118–1626
(540) 687–8390
www.awra.org
Contract: President

The AWRA is a nonprofit professional association founded in 1964 with a multidisciplinary membership of engineers, educators, foresters, biologists, ecologists, managers, and regulators. It offers a balanced, professional approach to solving water resource challenges, including research, establishment of a common meeting ground, and the collection and dissemination of ideas and information.

**American Wildlands**
PO Box 6669
40 East Main Street, Suite 2
Bozeman, MT 59771
(406) 586–8175
www.wildlands.org
Contact: Executive Director

For more than twenty-five years American Wildlands, a science-based regional conservation organization, has worked for on-the-ground change and has successfully led numerous wilderness and wild & scenic river initiatives throughout the American West.

**American Wind Energy Association**
1001 14th Street, NW, 12th Floor
Washington, DC 20005
www.awea.org

The 1,000 members of this trade association represent wind power-plant developers, wind turbine manufacturers, utilities, consultants, insurers, financial investors, researchers, and others who advocate on behalf of wind energy projects around the world.

**American Wood Council**
1111 Nineteenth Street, NW, Suite 800
Washington, DC 20036
(202) 463–2766
www.awc.org
Contact: President

Part of the American Forest & Paper Association (formerly the National Forest Products Association), the AWC is a trade group whose mission is to increase the use of wood by ensuring the broad regulatory acceptance of wood products, and influencing public policies affecting the use of wood products.

**Ancient Forest International**
PO Box 1850
Redway, CA 95560
(707) 923–3015
www.ancientforest.org
Contact: Director

This conservation alliance has conducted protection and preservation projects in several countries, including Mexico, Ecuador, Chile, and in the ancient forests of the United States. It fosters awareness and study of critical forest ecosystems and habitats.

**Appalachian Trail Conservancy**
799 Washington Street
PO Box 807
Harpers Ferry, WV 25425–0807
(304) 535–6331
www.appalachiantrail.org
Contact: Staff

As a volunteer-based, private nonprofit organization, the ATC is dedicated to conserving the Appalachian National Scenic Trail. The organization's mission is to ensure that future generations enjoy the clean air and water, scenic vistas, wildlife, and opportunities for recreation along the Appalachian Trail corridor.

**Argentina People and Nature Foundation**
668 San Martin Avenue
Huerta Grande, Cordoba 5174
Argentina
www.argentinapeopleandnature.org
Contact: President

To safeguard the cultural and natural heritage of Argentina, this group promotes the exchange of information, technology, and financial support for natural resource protection for future generations.

**BIOCAPE**
Kumaraperumal Vilai
South Tamarai Kulam, K.K. District
Tamil Nadu 629502
India
Contact: Secretary

The Biologists Association for Conservation and Public Education has as its motto "Heal the World." It is a nonprofit organization in India that works to protect natural resources by establishing rules for environmental protection and public education.

**Blue Ribbon Coalition**
1540 N. Arthur
Pocatello, ID 83204
(208) 233–6570
www.sharetrails.org
Contact: Executive Director

The group, founded in 1987, opposes the designation of additional protected wilderness areas, and is committed to defending the right to use public lands for recreation.

**Bluewater Network**
300 Broadway, Suite 28
San Francisco, CA 94133
(415) 788–3666
www.bluewaternetwork.org
Contact: Director

Working in opposition to the use of personal watercraft and other two-stroke engine-motored snowmobiles and off-highway vehicles, this organization attempts to keep motorized vehicles out of protected areas such as national parks.

**Boreal Forest Network**
3rd Floor, 303 Portage Avenue
Winnipeg, Manitoba R3B 1E7
Canada
(204) 947–3081
www.borealnet.org
Contact: Director

BFN is the North American affiliate of the Taiga Rescue Network. By working with indigenous people living in the Boreal Forest of North America, the organization's aim is to protect primary intact boreal forests.

**California Wilderness Coalition**
1212 Broadway, Suite 1700
Oakland, CA 94612
(510) 451–1450
Contact: Executive Director

The California Wilderness Coalition is the only organization dedicated to protecting California's wild places and native biodiversity on a statewide level. Through advocacy and public education, CWC builds support for threatened wild places, from oak woodlands to ancient forests and deserts, coordinating its efforts with community leaders, businesses, local organizations, and policy-makers.

**Ceiba Foundation for Tropical Conservation**
2319 N. Cleveland
Chicago, IL 60614
(773) 871–3798
www.ceiba.org
Contact: Director

Ceiba, a name taken from a giant tree found in tropical forests, was founded in 1997 to protect tropical habitats and to conserve their plants and animals in South and Central America. They sponsor scientific research, provide public education, and emphasize community-based projects to empower landowners and to encourage them to participate actively in sustainable land management.

**Center for Biological Diversity**
PO Box 710
Tucson, AZ 85702–0710
(520) 623–5252
www.biologicaldiversity.org
Contact: Executive Director

Although known primarily for its efforts to protect threatened and endangered species, the Center for Biological Diversity (CBD) considers the monitoring of public lands grazing one of its top priorities because of its impact on wildlife. The group opposes what it calls corporate welfare for the ranching industry —subsidies given to livestock owners that allow them to graze on public lands at rates considerably below the fees charged for private land grazing.

**Center for Science in Public Participation**
224 North Church Avenue
Bozeman, MT 59715
(406) 585–9854
www.csp2.org
Contact: Director

The center provides technical support, resources, and advice to grassroots organizations, nongovernmental organizations, regulatory agencies, businesses, and indigenous communities on natural resource issues, especially those relating to mining. Its goal is to give citizens information so they can make informed and proactive decisions on natural resource protection and development issues, and to ensure that extractive industry practices follow the highest standards.

**Coal Operators and Associates**
PO Box 3158
Pikeville, KY 41502
(606) 432–2161
www.miningusa.com/coa
Contact: President

Working for the welfare of the coal industry, the organization serves as a source of information for owners and operators, and provides representation in policy-making for its members.

**Colorado Environmental Coalition**
1536 Wynkoop Street # 5C
Denver, CO 80202
(305) 534–7066
www.ourcolorado.org
Contact: Executive Director

The Colorado Environmental Coalition's goal is to protect the state's natural heritage and quality of life. Every year, the coalition mobilizes thousands of organizations and individuals who care deeply about Colorado.

**Conservation Fund**
1655 N. Ft. Meyer Drive, Suite 1300
Arlington, VA 22209–2156
(703) 525–4610
www.conservationfund.org
Contact: Executive Director

Established in 1985, this organization works as a partner to public agencies and nonprofit groups to acquire property from landowners to provide protection to natural resources, wildlife habitat, historical sites, recreation areas, and open space.

**Conservation International**
1919 M Street, NW, Suite 600
Washington, DC 20036
(800) 406–2306
www.conservation.org
Contact: Chief Executive Officer

Although based in the United States as a nonprofit organization, Conservation International works in more than forty countries on four continents to preserve the earth's natural heritage. Its staff work in areas termed "biodiversity hot spots" in which animals, plants, and habitats are most at risk. These hot spots also include marine regions.

**Conservation Northwest**
1208 Bay Street #201
Bellingham, WA 98225–4301
(360) 671–9950
www.conservationnw.org
Contact: Executive Director

Conservation Northwest (formerly Northwest Ecosystem Alliance) was founded in 1988 and now has more than 8,000 members. In the last decade, Conservation Northwest has fought to maintain the ecological integrity of the Northwest's wildland, establishing itself as one of America's major regional conservation forces, combining organizing, mass media, and science skills with innovative strategy and fieldwork.

**Crude Accountability**
PO Box 2345

Alexandria, VA 22301
(703) 299–0854
www.crudeaccountability.org
Contact: Staff

The U.S.-based organization works with grassroots activists in the Caspian region to ensure environmental justice for local communities affected by oil development. Its focus is on bringing skills, training, and resources to small communities that have little access to the global environmental movement.

**Diamond High Council**
Hovenierstraat 22
BE–2018 Antwerp Belgium
www.hrd.be
Contact: Director

This industry organization provides training and certification on diamond grading, gemstone reports, research, and policy direction.

**Domestic Petroleum Council**
101 Constitution Avenue, NW, Suite 800
Washington, DC 20001–2133
(202) 742–4300
www.dpcusa.org
Contact: President

The nation's twenty-four largest independent natural gas and crude oil exploration and production companies are represented by this trade association. The group encourages responsible exploration, development, and production to meet consumer needs, and are leaders in developing technology for offshore production.

**Earth Island Institute**
300 Broadway, Suite 28
San Francisco, CA 94133
(415) 788–3666
www.earthisland.org
Contact: Executive Director

Founded by noted conservation activist David Brower after he left the Sierra Club, the institute promotes citizen action through a worldwide network of projects, from logging alerts and endangered species trafficking to encouraging public participation in environmental activism.

**Earthjustice Legal Defense Fund**
426 17th Street, Sixth Floor
Oakland, CA 94612–2820
(510) 550–6700
www.earthjustice.org
Contact: Executive Director

Earthjustice is the nonprofit law firm for the environment, representing—without charge—hundreds of public interest clients, large and small. Earthjustice works through the courts to safeguard public lands, national forests, parks, and wilderness areas; to reduce air and water pollution; to prevent toxic contamination; and to preserve endangered species and wildlife habitat.

**Earthstewards Network**
Box 10697
Bainbridge Island, WA 98110
(206) 842–7986
www.earthstewards.org
Contact: Program Director

This international nongovernmental organization is dedicated to providing the resources for conflict resolution of environmental disputes on a global level. It works on issues such as forest protection, urban forestry, and the rights of indigenous peoples to forest resources.

**Earthworks**
1612 K Street, NW, Suite 808
Washington, DC 20006
(202) 887–1872
www.mineralpolicy.org
Contact: Staff

This organization focuses on the impacts of mining and oil and gas development both in the United States and on an international level. Whether protesting cyanide contamination resulting from the processing of gold or efforts to expand corporate ownership of public lands, Earthworks monitors and publicizes activities it perceives as detrimental uses of natural resources.

**Ecological Society of America**
1707 H Street, NW, Suite 400
Washington, DC 20006
(202) 833–8773
www.esa.org
Contact: President

This nonpartisan nonprofit organization was founded in 1915 to promote ecological science and infuse ecological knowledge into environmental decision-making. Its 9,000 members work to support the Sustainable Biosphere Initiative, founded in 1992, to work with governmental agencies and the private sector in natural resource management, restoration, and sustainability.

**Energy Future Coalition**
1225 Connecticut Avenue, NW, Suite 400
Washington, DC 20036
(202) 463–1947
www.energyfuturecoalition.org
Contact: Executive Director

The coalition works to reduce the world's dependence on oil through a partnership of stakeholders representing business, labor, and nongovernmental organizations to take practical steps to replace petroleum and gas used in the production of fuels, chemicals, and plastics.

**Environmental Defense**
257 Park Avenue, South
New York, NY 10010
(212) 505–2100
www.environmentaldefense.org
Contact: Staff

By linking science, economics, and law, Environmental Defense works to protect the environmental rights of citizens to clean air, water, safe food, and a flourishing ecosystem.

**Environmental Working Group**
1436 U Street, NW, Suite 100
Washington, DC 20009
(202) 667–6982
www.ewg.org
Contact: Director

Since 1993, the organization has consisted of a team of scientists, engineers, policy experts, lawyers, and computer analysts who review government reports and data relating to science and regulations.

**Forest Action Network**
Box 625
Bella Coola
Nuxalk Territory
British Columbia Canada V0T 1C0
www.fanweb.org
Contact: Staff

FAN is a network of activists working to protect the forested coast of British Columbia, Canada, especially the Great Bear Rainforest. The group coalesced in 1993 with a direct action protest against road building in a pristine valley, and since then has attempted to stop raw log exports. An emphasis has now been placed on working with leaders of a dozen First Nations.

**Forest Guardians**
312 Montezuma
Santa Fe, NM 87501
(505) 988–9126
www.fguardians.org
Contact: Executive Director

Often considered one of the most contentious environmental organizations in the United States, Forest Guardians focuses on the preservation and restoration of native wildlands and wildlife in the American Southwest. Specific programs address the issues of forests, river restoration, endangered species, deserts and grasslands, and grazing reform.

**Forest Guild**
PO Box 519
Santa Fe, NM 87504
(505) 983–8992
www.foreststewardsguild.org
Contact: Director

A passion for forests and land stewardship is the common bond of the members of this nonprofit organization. Its aim is to promote economically, ecologically, socially responsible forestry as a way of maintaining the integrity of forest ecosystems and the human communities dependent upon them. It provides training, policy analysis, and research for its members in the United States and Canada.

**Forest Service Employees for Environmental Ethics**
PO Box 11615
Eugene, OR 97440
(541) 484–3170
www.fseee.org
Contact: Executive Director

The government employees and concerned citizens of FSEEE focus on holding the U.S. Forest Service accountable to the public, working toward an ecologically and economically sustainable value system for the nation's land management.

**Forests and the European Union Resource Network**
1C Fosseway Business Center
Stratford Road, Moreton-in-Marsh
Gloustershire GL56 9NQ
United Kingdom
www.fern.org
Contact: Staff

FERN works to create changes in policies and practices that impact forests and forest peoples' rights globally. It focuses on the European Union's policies dealing with the underlying causes of forest loss, conducting research, analysis, facilitation, coordination, support, education, and advocacy.

**Forests Forever**
50 First Street, Suite 401
San Francisco, CA 94105
(415) 974–3636
www.forestsforeever.org
Contact: Executive Director

The organization's major campaign is to save California's national forest lands by opposing logging. Its strategies include education outreach, legislative monitoring and lobbying, and electoral activities.

**Freshwater Action Network**
Prince Consort House, Seventh Floor
27–29 Albert Embankment
London SE1 7UB
United Kingdom
www.freshwateraction.net
Contact: Director

The purpose of the network is to assist community-based organizations to strengthen civil society participation in international water policy formulation. It works with a global network of nongovernmental organizations providing training and advocacy.

**Friends of the Environment for Development and Sustainability**
56-B Quezon Street, Don Domingo Maddela
Bayombong, Nueva Vizcaya
Philippines
Contact: Staff

FRIENDS, formed in 1999, works to implement conservation activities in the Palali-mamparang Mountain Range in the province. It also deals with environmental management within the Sierra Madre Biodiversity Corridor, part of the Philippine Biodiversity Conservation Priority Project.

**Fundacion Ambiente y Recursos Naturales**
Monroe 2142 1 B (1428)
Buenos Aires, Argentina
www.farn.org.ar
Contact: Director

FARN is a nonprofit organization whose mission is to promote sustainable development through policy, law, and institutional organization. Through projects such as Sea and Sky, the group coordinates conferences and workshops on natural resources and development.

**Gas Processors Suppliers Association**
6526 E. 60th Street
Tulsa, OK 74145
(918) 493–3872

www.gasprocessors.com
Contact: President

As a forum for the exchange of information about global gas processing, this group expands technical knowledge, provides information about best practices for the industry, analyzes legislation and regulations, and works to enhance perspectives on the market.

**Global Forest Watch**
10 G Street, NW
Washington, DC 20002
(202) 729–7600
www.globalforestwatch.org
Contact: Director

An initiative of the World Resources Institute, this project provides technical assistance and support for preserving forest ecosystems around the world. One of its primary resources is the use of global satellite imagery that assists planners and policymakers in identifying current forest cover and remaining frontier forests, especially in remote areas.

**Global Policy Forum**
777 United Nations Plaza, Suite 3D
New York, NY 10017
(212) 557–3161
www.globalpolicy.org
Contact: Executive Director

This nongovernmental organization monitors policy-making at the United Nations, promotes accountability of global decisions, educates and mobilizes for global citizen participation, and advocates on vital issues of international peace and justice.

**Global Water**
3600 S. Harbor Boulevard #514
Oxnard, CA 93035
(805) 985–3057
www.globalwater.org
Contact: Staff

By helping people to help themselves, the organization links the need for adequate supplies of water with disease, hunger, and poverty on a global level.

**Global Water Research Coalition**
Alliance House
12 Caxton Street
London SW1H 0QS
United Kingdom
www.globalwaterresearchcoalition.net
Contact: Director

As its name implies, this group promotes international cooperation and collaboration on water issues. The coalition was organized in 2002 in association with numerous nongovernmental associations and researchers to disseminate information and
sponsor meetings.

**Grazing Lands Conservation Initiative**
501 W. Felix Street, Building 23
Ft. Worth, TX 76115
(817) 509–3318
www.glci.org
Contact: National Coordinator

Established in 1991, this group seeks to provide technical assistance to livestock owners on privately owned land to increase the
awareness of grazing land resources. It implements its mission
through coalitions at the local, state, and national levels including
livestock owners, scientists, environmental groups, and state and
federal agencies.

**Great Old Broads for Wilderness**
PO Box 2924
Durango, CO 81302
(970) 385–8303
www.greatoldbroads.org
Contact: Staff

Although this group's goal is to preserve the wilderness for future generations, its activities center around livestock grazing, encouraging the voluntary buyout and retirement of grazing permits to potentially heal public lands that have been degraded.
They support the use of sound scientific research to guide public
agencies in enforcing grazing management plans.

**Greater Yellowstone Coalition**
PO Box 1874
Bozeman, MT 59771
(406) 686–1593
www.greateryellowstone.org
Contact: Executive Director

The Greater Yellowstone Coalition (GYC) focuses on protecting the land, water, and wildlife of the Greater Yellowstone Ecosystem. GYC was founded in 1983 on the premise that an ecosystem will remain healthy and wild only if it is kept whole. Since that time the organization has emerged as a nationally known advocate for the idea that ecosystem-level sustainability should guide the management of the region's public and private lands.

**Heritage Forests Campaign**
1200 Eighteenth Street, NW, Fifth Floor
Washington, DC 20036
(202) 887–8800
www.ourforests.org
Contact: Executive Director

The Heritage Forests Campaign is an alliance of conservationists, wildlife advocates, clergy, educators, scientists, and other Americans working together to uphold protection of the national forests. Heritage Forests Campaign's partners include the Alaska Rainforest Coalition, American Hiking Society, Earthjustice, National Environmental Trust, National Audubon Society, Natural Resources Defense Council, U.S. PIRG, and the Wilderness Society.

**High Country Citizens' Alliance**
PO Box 1066
Crested Butte, CO 81224
(970) 349–7104
www.hcaonline.org
Contact: Executive Director

This group works with stakeholders to improve rangeland health, seeking agency enforcement of existing grazing standards, encouraging education, supporting intensive management and voluntary retirement of grazing allotments.

**Idaho Conservation League**
PO Box 844

Boise, ID 83701
(208) 345–6933
www.wildidaho.org
Contact: Program Director

The Idaho Conservation League seeks to preserve Idaho's clean water, wilderness, and quality of life through citizen action, public education, and professional advocacy.

**Independent Petroleum Association of America**
1201 Fifteenth Street, NW
Washington, DC 20005
(202) 857–4722
www.ipaa.org
Contact: Director

Thousands of independent oil and natural gas producers are represented by this trade association, which was founded in 1929 in Colorado. They represent 85 percent of the wells in the United States and produce 65 percent of the country's natural gas and 40 percent of its oil. The group serves as a national advocate for its members and compiles information about the domestic exploration and production industry.

**Institute for Problems of Natural Resource Use and Ecology**
10 Starabarysauski Trakt
Minsk BY–220114
Republic of Belarus
www.ecology.ac.by
Contact: Director

Founded in 1932, this research institute develops ecologically safe and resource-saving technologies for mining and processing.

**Instituto Terra**
Fazenda Bulcao-CX Postal–005
CEP 35200–000
Almores, MG, Brazil
www.institutoterra.org
Contact: Director

Created in 1998 in the Rio Doce Valley of Brazil, the institute is committed to creating long-term environmental and social change to return the land and resources to its once-rich diversity.

It is a leading center for environmental education, sustainable development, and social mobilization in Brazil.

**International Association of Oil and Gas Producers**
209–215 Blackfriars Road
London SE1 8NL
United Kingdom
www.ogp.org.uk
Contact: Director

This international trade association represents the producers of more than half of the world's oil and about one-third of global natural gas production.

**International Gas Union**
PO Box 550
Agern Alle 24–26
2970 Hoersholm Denmark
www.igu.org
Contact: Director

Founded in 1931, the IGU is registered in Switzerland and operates its secretariat in Denmark. The organization promotes the technical and economic progress of the natural gas industry through associations in sixty-seven countries, and it deals with the exploration and production of gas, gas storage, distribution and transmission, sustainable development, and developing markets for natural gas.

**International Tropical Timber Organization**
International Organizations Center, 5th Floor
1–1–1 Minato-Mirai, Nishi-ku
Yokahama, 220–0012 Japan
www.itto.org
Contact: Director

Nearly fifty governments participate in the activities of the ITTO, which works to formulate and implement various treaties and agreements among countries that produce or import tropical timber.

**International Water Resources Association**
4535 Faner Hall, MC 4516
Southern Illinois University

Carbondale, IL 62901–4516
(618) 453–5138
www.iwra.siu.edu/worldwater
Contact: Executive Director

The IWRI is a nonprofit, nongovernmental organization hosted by Southern Illinois University to promote sustained water use on an international scale. More than 110 countries are represented as part of the group's efforts to promote water resources education and networking, and to provide an international forum on water resources.

**Izaak Walton League**
707 Conservation Lane
Gaithersburg, MD 20878
(301) 548–0150
www.iwla.org
Contact: Executive Director

The 50,000 members of the IWL belong to one of the nation's oldest conservation organizations. Its grassroots approach to the sustainability of natural resources brings together hunters, photographers, anglers, birders, and other outdoor enthusiasts.

**Jewelers Vigilance Committee**
25 West 45th Street, Suite 1406
New York, NY 10036
(212) 997–2002
www.jvclegal.org
Contact: President

As a nonprofit trade association representing the jewelry industry, this organization, founded in 1912, serves as a resource for consumers and dealers, providing compliance information relating to diamond issues.

**Lifewater International**
3765 S. Higuera Street #120
San Luis Obispo, CA 93401
(805) 541–6634
www.lifewater.org
Contact: Staff

Since 1979, the volunteers and trainers in this nongovernmental organization have served more than 1 million people in developing countries by teaching local communities how to improve their drinking water systems in areas in which clean water is a scarce resource.

**Lignite Energy Council**
1016 Owens Avenue
PO Box 2277
Bismarck, ND 58502
(701) 258–7177
www.lignite.com
Contact: Director

The LEC represents coal and natural gas producers in an effort to enhance resources and mining by-products for use in electricity generation in the United States.

**Mineral Information Institute**
501 Violet Street
Golden, CO 80401
(303) 277–9190
www.mii.org
Contact: Director

The MII is focused on educational programs relating to natural resources that provide an understanding of the role of minerals and energy in society, and how they can be produced in an environmentally and socially responsible way.

**National Association of State Foresters**
444 N. Capitol Street, NW, Suite 540
Washington, DC 20001
(202) 624–5415
www.stateforesters.org
Contact: Executive Director

Working with nonindustrial private forest landowners, this group provides technical assistance, develops outreach and education programs, and assists in urban and community forestry.

**National Audubon Society**
700 Broadway
New York, NY 10003
(212) 979–3000

www.audubon.org
Contact: Executive Director

Audubon's mission is to conserve and restore natural ecosystems, with a focus on birds, wildlife, and their habitats for the benefit of humanity and the earth's biological diversity. The group's national network of community-based nature centers and chapters and scientific and educational programs provide advocacy in behalf of areas sustaining important bird populations.

**National Cattlemen's Beef Association**
444 North Michigan Avenue, Suite 1800
Chicago, IL 60611
(312) 467–5520
www.beefusa.org
Contact: Executive Director

Organized in 1898, this is one of the oldest and most established industry groups, representing more than 1 million cattle farmers and ranchers. It seeks a profitable beef industry through state beef councils, an extensive research program, and political advocacy to advance its members' interests.

**National Commission on Science for Sustainable Forestry**
1707 H Street, NW, Suite 200
Washington, DC 20006
(202) 207–0007
www.ncseonline.org
Contact: Director

The NCSSF's mission is to improve the research and scientific basis relating to sustainable forestry. It sponsors workshops and symposia but maintains a position of nonadvocacy in representing forest managers.

**National Energy Foundation**
3676 California Avenue, Suite A117
Salt Lake City, UT 84104
(800) 616–8326
www.nef1.org
Contact: President

The foundation is a tax-exempt, publicly supported education group that provides curriculum, training, and other support to facilitate an understanding of energy, natural resources, and the environment.

**National Environmental Trust**
1200 Eighteenth Street, NW, Fifth Floor
Washington, DC 20036
(202) 887–8800
www.net.org
Contact: President

The National Environmental Trust (NET) is a nonprofit, nonpartisan organization established in 1994 to inform citizens about environmental problems and how they affect health and quality of life. NET's public education campaigns use modern communication techniques and the latest scientific studies to translate complex environmental issues for citizens to localize the impacts of national problems, as well as to highlight opportunities for Americans to engage in the policy-making process.

**National Forest Foundation**
Building 27, Suite #3
Fort Missoula Road
Missoula, MT 59804
(406) 542–2805
www.natlforests.org
Contact: Director

This group's focus on building partnerships to protect and restore grasslands and national forests makes it both an advocate for change and a resource for other organizations.

**National Mining Association**
101 Constitution Avenue, NW
Suite 500 East
Washington, DC 20001-2133
(202) 463–2600
www.nma.org
Contact: President

Representing more than 325 corporations involved in the mining industry, this nonprofit group was formed in 1995 with the merger of the National Coal Association and the American Mining Congress. Its purpose is to develop political support for the mining industry to utilize the nation's mineral resources.

**National Park Trust**
51 Monroe Street, Suite 110
Rockville, MD 20850
(301) 279–PARK
www.parktrust.org
Contact: Director

This private citizen organization works in cooperation with the National Park Service to protect the resources within and around national parks, historic sites, and natural areas.

**National Parks Conservation Association**
1300 Nineteenth Street, NW, Suite 300
Washington, DC 20036
(800) 628–7275
www.npca.org
Contact: President

One of the nation's oldest and largest conservation organizations, the association is dedicated to providing support for the national park system, advocating for designations, infrastructure develop-ment and maintenance, preservation, and lobbying members of Congress.

**National Public Lands Grazing Campaign**
1213 Iowa Street
Ashland, OR 97520
(541) 201–0053
www.publiclandsranching.org
Contact: Executive Director

This organization has served at the forefront of efforts to end live-stock grazing on public lands by educating the public about how livestock affect the ecological, economic, and fiscal environment, seeking enforcement of existing grazing management laws, and calling for the amendment of federal law to allow the voluntary buyout of grazing permits.

**National Wildlife Federation**
11100 Wildlife Center Drive
Reston, VA 20190–5362
(703) 438–6000
www.nwf.org
Contact: President

Founded in 1936, the NWF is one of the early mainstream environmental organizations in the United States. Over the years it has expanded its agenda from protection of species and habitats to dozens of programs in forty-seven states and territories, including efforts to stop oil development in the Arctic National Wildlife Refuge and other public lands.

**Native Forest Network**
PO Box 8251
Missoula, MT 59807
(406) 251–2385
www.nativeforest.org
Contact: Director

The NFN is a collaborative partnership of organizations working to protect native forests throughout the world. It serves as an advocacy group, and provides resources for organizations to help them promote forest resources.

**Native Habitat**
PO Box 100671
Ft. Worth, TX 70185
www.nativehabitat.org
Contact: President

Composed of livestock managers, students, environmental group members, wildlife biologists, and scientists, this grassroots group supports the development of healthy rangelands to protect natural resources from degradation through sustainable practices.

**Natural Resources Defense Council**
40 W. Twentieth Street
New York, NY 10011
(212) 727–2700
www.nrdc.org
Contact: Executive Director

NRDC is considered by many to be the nation's most effective environmental action organization. The group uses law, science, and the support of more than 1 million members and online activists to protect the planet's wildlife and wild places and to ensure a safe and healthy environment for all living things.

**Nature Conservancy**
4245 N. Fairfax Drive, Suite 100
Arlington, VA 22203
(877) 812–3698
www.nature.org
Contact: Chairman, Board of Directors

TNC's mission is to preserve the plants, animals, and natural communities that represent the diversity of life on earth by protecting the lands and waters they need to survive. The primary mechanism for doing so is through community engagement.

**New Forests Project**
731 Eighth Street, SE
Washington, DC 20003
(202) 547–3800
www.newforestsproject.com
Contact: Director

By stressing the importance of trees to the Earth's ecosystems, the program provides information about forest resources, seeks to provide rural communities in Central America with clean drinking water, and sets up seed banks around the world.

**Oil & Gas Accountability Project**
863 Main Avenue
Durango, CO 81301
(970) 259–3353
www.ogap.org
Contact: Executive Director

This organization works with communities across the Rocky Mountain West to reduce the social, economic, and environmental problems caused by oil and gas development. With offices in both Colorado and Montana, OGAP works as a watchdog citizens' group to inform communities about toxic chemicals associated with industry activities.

**Oregon Natural Desert Association**
16 Kansas Avenue
Bend, OR 97701
(541) 330–2638

www.onda.org
Contact: Executive Director

One of the few organizations devoted to the protection of native deserts, this group was organized in the mid-1980s to oppose the Bureau of Land Management's plans to designate wilderness study areas in eastern Oregon. A nonprofit organization, the association monitors livestock grazing, mining, and geothermal development.

**Oregon Natural Resources Council**
5825 North Greeley Avenue
Portland, OR 97217–4145
(503) 283–6343
www.onrc.org
Contact: Executive Director

Founded in 1974, the ONRC has a mission of aggressively protecting and restoring Oregon's wildlands, wildlife, and waters as an enduring legacy. ONRC has been instrumental in securing permanent legislative protection for some of Oregon's most precious landscapes, including nearly 1.5 million acres of wilderness, 1,700 miles of wild and scenic rivers, and more than 58 million acres of roadless areas across the country.

**Pacific Institute for Studies in Development, Environment, and Security**
654 13th Street, Preservation Park
Oakland, CA 94612
(510) 251–1600
www.pacinst.org
Contact: President

Founded in 1987, the institute hosts a website that provides a chronology describing ways in which water has been fought over, used as a weapon, or targeted in violent conflict.

**Partners in Parks**
PO Box 130
Paonia, CO 81428
(970) 527-6675
www.partnersinparks.org
Contact: Staff

By forging relationships with professionals involved in public land management, this organization focuses on the study and interpretation of natural resources in cooperation with the National Park Service.

**Partnership Africa Canada**
323 Chapel Street
Ottawa, Ontario
K1N 7Z2 Canada
(613) 237–6768
www.pacweb.org
Contact: Executive Director

The partnership was created in 1986 to work with organizations in Africa, Canada, and internationally to promote sustainable use of natural resources and development in Africa. It funds innovative projects and conducts policy research relating to human rights and human security, including the diamond trade.

**Pinchot Institute for Conservation**
1616 Park Street, NW
Washington, DC 20036
(202) 797–6580
www.pinchot.org
Contact: Director

Named after the first chief of the U.S. Forest Service, the Pinchot Institute brings together the nation's leading land managers, foresters, conservationists, and researchers to forge new forest policy directions.

**Public Employees for Environmental Responsibility**
2001 S Street, NW, Suite 570
Washington, DC 20009
(202) 265–4192
www.peer.org
Contact: Executive Director

Local, state, and federal employees working for resource management agencies are represented in this national, nonprofit organization that serves as a watchdog for the public. It supports public employees who speak out about issues relating to natural resource management, and provides legal representation to whistle blowers bringing attention to environmental enforcement.

**Quincy Library Group**
159 Lawrence Street
PO Box 11500
Quincy, CA 95971
(530) 283–7769
www.qlg.org
Contact: Project Manager

The timber wars of the late 1980s and early 1990s were the impetus behind the formation of the Quincy Library Group, founded in 1992. The organization focuses on three counties in northeastern California, and began with recommendations on how best to manage the national forests while maintaining community stability.

**Rainforest Foundation USA**
32 Broadway, Suite 1614
New York, NY 10004
(212) 431–9098
www.rainforestfoundation.org
Contact: Director

Based in New York, the Rainforest Foundation has sister organizations in the United Kingdom and in Norway, dedicated to protecting the rights of rain forest groups in their efforts to protect their environment. Founded in 1989, RF assists local grassroots organizations with current projects in Colombia, Guyana, Suriname, Brazil, and Ecuador.

**RangeNet**
2850 SW Cedar Hills Boulevard
Beaverton, OR 97005
www.rangenet.org
Contact: Staff

This network of individuals working to improve the ecological conditions of America's public rangelands is associated with the Western Watersheds Project, based in Hailey, Idaho. RangeNet serves as a communications and information link for rangeland activists; membership is available through nomination and invitation.

**Resource Renewal Institute**
Fort Mason Center, Building D
San Francisco, CA 94123
(415) 928–3774

www.rri.org
Contact: Executive Director

The institute develops innovative solutions for long-term conservation and a sustainable future through programs such as Defense of Place, designed to create a principle in the minds of Americans that protected lands will be protected forever.

**Rocky Mountain Mineral Law Foundation**
9191 Sheridan Boulevard, Suite 203
Westminster, CO 80031
(303) 321–8100
www.rmmlf.org
Contact: Staff

Founded in 1955 as an educational organization dedicated to providing scholarly research on the laws and issues affecting both domestic and global minerals and water resources, the foundation works with law schools, oil and gas companies, mining interests, water and public land users, conservation groups, and representatives of environmental protection agencies. It sponsors short courses and publications by volunteers with specific expertise in natural resource law.

**Saguaro Juniper Corporation**
5707 E. Sixth Street
Tucson, AZ 85711
(520) 745–6025
www.saguaro-juniper.com
Contact: Staff

Founded in 1988, this small group of Arizona cattle ranchers pooled their financial resources to purchase land along the San Pedro River where they practice the protection of riparian areas, grazing on a rotational basis, the humane treatment of animals, and the resting of the land.

**San Juan Citizens Alliance**
PO Box 2461
1022½ Main Avenue
Durango, CO 81302
(970) 259–3583
www.sanjuancitizens.org
Contact: Executive Director

With offices in both Colorado and New Mexico, the 500-member group, founded in 1984, was established to provide a voice for environmental, economic, and social justice in the region. The group is engaged in an effort to restore more than 2 million acres of wild habitat in the San Juan Mountains, is opposing efforts for new gas wells in the area near Farmington, New Mexico, and has successfully opposed development of the last wild valley in the San Juans.

**Seacology**
2009 Hopkins Street
Berkeley, CA 94707
(510) 559–3505
www.seacology.org
Contact: Staff

The staff and members of Seacology are dedicated to preserving the endangered natural resources and biodiversity of the world's islands. An effort is made to conduct community-based projects such as building schools in exchange for protection of the local environment.

**Sierra Club**
85 Second Street, Second Floor
San Francisco, CA 94105
(415) 977–5500
www.sierraclub.org
Contact: Executive Director

Founded in 1892, the Sierra Club promotes the protection of the nation's wilderness areas, including forests, national parks, recreation areas, and natural resources. It is one of the major environmental organizations in the United States, with a history of outreach and advocacy on key issues.

**Society for Range Management**
10030 West 27th Avenue
Wheat Ridge, CO 80215–6601
(303) 986–3309
www.rangelands.org
Contact: President

This membership-based organization develops policies and position statements on issues relating to rangeland assessment and monitoring, and provides training and technical assistance for professional certification and leadership development.

**Society of American Foresters**
5400 Grosvenor Lane
Bethesda, MD 20814–2198
(301) 897–8720
www.safnet.org
Contact: President

Gifford Pinchot founded the SAF in 1900 as part of an effort to professionalize forestry in the United States. Since that time the organization has worked to promote the conservation ethic through training, advocacy, outreach, and technology development to ensure the availability of forest resources for future generations.

**Soil and Water Conservation Society**
945 SW Ankeny Road
Ankeny, IA 50023–9723
(515) 289–2331
www.swcs.org
Contact: Executive Director

Through research, education, and advocacy, this group seeks to improve the practice of natural resource conservation, and represents the interests of ranchers, policy-makers, educators, planners, farmers, and students.

**Southeast Alaska Conservation Council**
419 Sixth Street, Suite 200
Juneau, AK 99801
(907) 586–6942
www.swcs.org
Contact: Executive Director

Founded in 1970, SEACC is a coalition of eighteen member groups in fourteen communities, stretching along the coast from Ketchikan to Yakutat, whose goal is to safeguard the sustainable use of the region's natural resources. Together with allied groups and individuals, the organization tries to reduce clear-cutting on the Tongass National Forest.

**Southern Appalachian Forest Coalition**
46 Haywood Street, Suite 323
Asheville, NC 28801
(828) 252–9223
www.safc.org
Contact: Staff

The Southern Appalachian Forest Coalition (SAFC) is a nonprofit organization formed in 1994 to address the increasing threats facing Southern Appalachian public lands. The objectives of the coalition are to create a unified and compelling regional conservation vision; to achieve greater representation in Washington, D.C.; and to strengthen grassroots groups with the tools and leadership needed to protect forests.

**Southern Pine Council**
2900 Indiana Avenue
Kenner, LA 70065–4464
(504) 443–4464
www.southernpine.com
Contact: Staff

The council is a joint promotional body coordinated and supported by manufacturers of southern pine lumber who are also members of the Southern Forest Products Association and Southeastern Lumber Manufacturers Association. The group works to provide product information, lumber applications, and technical assistance to domestic and international buyers.

**Southern Rockies Conservation Alliance**
1660 Wynkoop Street, Suite 850
Denver, CO 80202
(303) 650–5818
www.southernrockies.org
Contact: Staff

The SRCA is a coalition of conservation and recreation organizations focusing on the development and implementation of coordinated, comprehensive, and effective campaigns to protect wilderness quality lands, conserve and restore biodiversity, ensure responsible management of motorized recreation, promote healthy forest ecosystems, and restore natural fire regimes.

**Stewards of the Range**
PO Box 490
Meridian, ID 83680–0490
(208) 855–0707
www.stewards.us
Contact: Executive Director

Incorporated as a nonprofit organization, Stewards was created in 1992 to litigate on behalf of landowners and protect private property rights. It works with property owners, ranchers, and others seeking to challenge environmental regulations that result in a taking of private land or property without compensation.

**Trees for the Future**
PO Box 7027
Silver Spring, MD 20907
(301) 565–0630
www.treesftf.org
Contact: Staff

After land has been degraded, technicians from this organization provide assistance by introducing environmentally sustainable management projects, providing planting materials, and educating the public about deforestation in areas such as the Amazon.

**Tropical Forest Foundation**
2121 Eisenhower Avenue, Suite 200
Alexandria, VA 22314
(703) 518–8834
www.tropicalforestfoundation.org
Contact: Executive Director

The foundation is a nonprofit education institution that is widely recognized for developing model programs and teaching the principles of sustainable forest management. Its emphasis is on encouraging the major international timber companies to incorporate reduced-impact logging as a more environmentally sensitive alternative to conventional logging.

**Washington Wilderness Coalition**
123 NW 36th Street, Suite 220
Seattle, WA 98107

(206) 633–1992
www.wawild.org
Contact: Executive Director

The group works to preserve and restore wild areas in Washington State through citizen empowerment, support for grassroots community groups, advocacy, and public education.

**Western Clean Energy Campaign**
2260 Baseline Road, Suite 205
Boulder, CO 80302
(303) 440–7517
www.westernresources.org
Contact: Director

Formed to oppose two dozen coal-fired power plants proposed in the West, the group organizes local communities, builds coalitions, and uses the news media and legal challenges to stop power plant expansion.

**Western Forestry and Conservation Association**
4033 SW Canyon Road
Portland, OR 97221
(503) 226–4562
www.westernforestry.org
Contact: Staff

The association works with professional foresters providing continuing education workshops and seminars throughout the Pacific Northwest. It emphasizes the use of best practices for reforestation, management, and conservation.

**Western Land Exchange Project**
PO Box 95545
Seattle, WA 98145
(206) 325–3503
www.westlx.org
Contact: Staff

Research and outreach are the two primary strategies used by this group, which monitors federal land exchanges, conveyances, and sales so that the public interest is served and environmental laws are upheld in the eleven Western states. The organization advocates policy reform against agencies that undervalue public lands or fail to protect the environment, by applying pressure for reform.

**Western Resource Advocates**
2260 Baseline Road, Suite 200
Boulder, CO 80302
(303) 444–1188
www.westernresourceadvocates.org
Contact: Executive Director

Since 1989, the WRA has served as a nonprofit environmental law and policy organization dedicated to restoring and protecting the natural environment of the American West. The group works to develop strategic programs in three areas: water, energy, and lands, collaborating with other environmental and community groups, and by developing solutions that are appropriate to the unique environmental, economic, and cultural framework of the region.

**Western Wood Products Association**
522 SW Fifth Avenue, Suite 500
Portland, OR 97204–2122
(503) 224–3930
www.wwpa.org
Contact: President

This trade association represents softwood lumber manufacturers in the twelve Western states and Alaska. Members have access to information and technical assistance relating to timber production, quality control, international trade, and support for sawmill owners and lumber product users.

**Wild Wilderness**
248 NW Wilmington Avenue
Bend, OR 97701
(541) 385–5261
www.wildwilderness.org
Contact: Executive Director

In 1991, the organization started as local residents began participating in local Forest Service meetings and learned of a plan to turn a local winter recreation area into a snowmobile park. Since that time, the group has focused on opposition to the Forest Service's Fee Demonstration Program.

**Wilderness Society**
1615 M Street, NW
Washington, DC 20036
(800) THE–WILD
www.wilderness.org
Contact: Staff

The Wilderness Society, founded in 1935, advances its agenda through a combination of science, advocacy, and education. Their mission is to deliver to future generations an unspoiled legacy of wild places, focusing on biological diversity; clean air and water; and forests, rivers, and deserts. The staff brings together scientific expertise, analysis, and advocacy at the highest levels to save, protect, and restore America's wilderness areas.

**Wildlands Project**
PO Box 455
Richmond, VT 05477
(802) 434–4077
www.twp.org
Contact: Director

With the motto "Reconnect, Restore, Rewild," the Wildlands Project seeks to create linkages to tie natural places together using mosaics of public and private lands and voluntary partnerships.

**Winrock International**
2101 Riverfront Drive
Little Rock, AR 72202
(501) 280–3000
www.winrock.org
Contact: Staff

As a nonprofit organization, Winrock works with individuals and communities worldwide to match innovative approaches in agriculture, natural resources management, and clean energy with technology and ideas to benefit the poor and disadvantaged of the world.

**Woodland Trust**
Autumn Park, Dysart Road
Grantham, Lincolnshire NG31 6LL
United Kingdom

www.woodland-trust.org.uk
Contact: Director

The trust is the United Kingdom's leading conservation charity dedicated to the protection of the native woodland heritage. Its members are attempting to have the oak tree declared a national icon, and working toward reforestation and restoration on both the large and small scales.

**Woodrow Wilson International Center for Scholars:**
**Environmental Change and Security Project**
1 Woodrow Wilson Plaza
1300 Pennsylvania Avenue, NW
Washington, DC 20004-3027
(202) 691–4000
wwics.si.edu
Contact: Director

The Wilson Center studies the linkages among natural resources, conflict, human health, and foreign policy, and provides links to news, events, and publications, including scholarly research.

**World Conservation Union**
Rue Mauverney 28
Gland 1196
Switzerland
www.iucn.org
Contact: Director General

The World Conservation Union, also known as the International Union for the Conservation of Nature and Natural Resources, was founded in 1948 as a network of more than 1,000 representatives of national governments and agencies, nongovernmental organizations, and scientists and experts from 181 countries. Its mission is to encourage and assist stakeholders to make the use of natural resources equitable and ecologically sustainable.

**World Diamond Council**
52 Vanderbilt Avenue, 19th Floor
New York, NY 10017
www.worlddiamondcouncil.org
Contact: Chairman

Meeting in Antwerp, Belgium, in July 2000, the members of the International Diamond Manufacturers Association and the World Federation of Diamond Bourses created the WDC to track the export and import of rough diamonds.

**World Energy Council**
Regency House, Fifth Floor
1–4 Warwick Street
London W1B 5LT UK
www.worldenergy.org
Contact: Director

For more than eighty years, the WEC has represented multi-energy corporations in ninety countries, covering coal, oil, natural gas, nuclear, hydro, and renewable energy. The group is accredited by the United Nations as a nongovernmental, noncommercial, and nonaligned registered charity.

**World Gold Council**
45 Pall Mall
London SW1Y 5JG
United Kingdom
www.gold.org
Contact: Director

As an association limited to industry members, the World Gold Council represents large companies dealing with gold production in Australia, Canada, India, Japan, Peru, South Africa, Tanzania, and the United States.

**World Resources Institute**
10 G Street, NE
Washington, DC 20002
(202) 729–7600
www.wri.org
Contact: Executive Director

WRI describes itself as an environmental think tank whose mission is to move human society to live in ways that protect the earth's environment for current and future generations. Its goal is to reverse damage to ecosystems, expand participation in environmental decisions, avert dangerous climate change, and increase prosperity while improving the environment by building bridges between ideas and action. WRI was launched in 1982 with a \$15 million grant from the MacArthur Foundation.

**World Stewardship Institute**
409 Mendocino Avenue, Suite A
Santa Rosa, CA 95401–8513
(707) 573–3160
www.ecostewards.org
Contact: Director

The organization was established to cultivate stewardship on an international level, linking groups involved with cooperative problem solving. Its major programs deal with worldwide forestry, biodiversity, and a Latin American alliance.

**World Wildlife Fund**
Avenue du Mont Blanc 1196
Gland, Switzerland
www.panda.org
Contact: Director

The WWF's mission sounds simple: to stop the degradation of the planet's environment and to build a future in which humans live in harmony with nature. But since its founding in 1961, that mission has led to programs in 100 countries and 2,000 conservation projects with a staff of 4,000. Its field-based projects work with indigenous peoples and seek to avoid unnecessary confrontation.

# U.S. Government Agencies

**Bureau of Indian Affairs**
1849 C Street, NW
Washington, DC 20240
(202) 208–5116
www.doi.gov/bureau-indian-affairs

Under the Department of the Interior, the BIA is responsible for protection of tribal lands across the United States and the conservation of resources on those lands. The agency works with Native American leaders to retain cultural ties to the land and the management of natural resources.

**Bureau of Land Management**
1849 C Street, NW
Washington, DC 20240
(202) 452–5125
www.blm.gov
Contact: Director

Under the Department of the Interior, the BLM administers 262 million acres of public lands, primarily in twelve Western states, representing one-eighth of the land surface in the United States. Programs cover energy and mineral resources, wilderness areas, fish and wildlife habitat, wild horses and burros, and natural resource development and protection.

**Cooperative State Research, Education, and Extension Service**
1400 Independence Avenue, SW
Washington, DC 20250–2201
(202) 720–7441
www.reeusda.gov
Contact: Director

The Cooperative Extension Service supports university-based research and leadership programs focusing on timber and logging practices. It offers education programs and technical information through field offices covering the entire United States.

**Forest Service**
PO Box 96090
Washington, DC 20090–6090
(202) 205–8333
www.fs.us.gov
Contact: Chief

Founded by Gifford Pinchot, the nation's first professional forester, the agency manages the national forests and grasslands through ranger districts and regional offices. Its primary mission is to ensure the productivity of forests on public lands, and to provide for sustainable management for current and future generations.

**Geological Survey**
12201 Sunrise Valley Drive
Reston, VA 20192
(888) 275–8747
www.usgs.gov
Contact: Director

With a staff of over 10,000 scientists, technicians, and support employees, the USGS has more than 400 offices throughout the United States. Its purpose is to provide scientific information and research relating to the nation's natural resources and scientific phenomena. The agency was established in 1879.

**Minerals Management Service**
1849 C Street, NW
Washington, DC 20240
(202) 208–3985
www.mms.gov
Contact: Chief, Public Affairs

The MMS is a bureau in the Department of the Interior responsible for managing the nation's natural gas, oil, and other mineral resources. It also collects, monitors, and disburses revenues from federal offshore mineral leases and from onshore mineral leases on Indian lands.

**National Park Service**
1849 C Street, NW
Washington, DC 20240
(202) 208–6843
www.nps.gov
Contact: Director

As part of the U.S. Department of the Interior, the National Park Service has had oversight of the nation's cultural and natural resources since its founding in 1916. The agency holds responsibility for a variety of sites including national monuments, parkways, battlefields, and historic sites.

**National Petroleum Council**
1625 K Street, NW, Suite 600
Washington, DC 20006
(202) 393–6100

www.npc.org
Contact: Executive Director

Established in 1946, the NPC acts as an oil and natural gas advisory committee to the secretary of the Department of Energy. The council's sole purpose is to represent the views of the oil and gas industries in advising, informing, and making recommendations on any matter relating to oil and natural gas.

**Natural Resources Conservation Service**
PO Box 2890
Washington, DC 20013
(202) 720–7246
www.nrcs.usda.gov
Contact: Executive Director

Originally called the Soil Conservation Service, this federal agency provides technical assistance and leadership to land users, communities, state and local governments, and other federal agencies planning and implementing conservation systems.

**Office of Fossil Energy**
1000 Independence Avenue, SW
Washington, DC 20585
(800) DIAL–DOE
www.fe.doe.gov
Contact: Director

As part of the Department of Energy, this bureau deals with federal regulations for clean coal, natural gas, hydrogen and other clean fuels, oil, and strategic petroleum reserves. It provides funding for innovative projects, technical assistance, and provides research information and project data.

**Office of Surface Mining**
1951 Constitution Avenue, NW
Washington, DC 20240
(202) 208–2719
www.osmre.gov
Contact: Director

An agency under the Department of the Interior, the OSM implements the Surface Mining Law, regulates active and abandoned mines, conducts research on reclamation and technology, and collects fees from industries engaged in surface mining activities.

**U.S. Department of Agriculture**
1400 Independence Avenue, SW
Washington, DC 20250
(202) 720–3631
www.usda.gov
Contact: Secretary

This agency includes several resource divisions, including the Bureau of Land Management, Cooperative Extensive Service, Natural Resource Conservation Service, and Forest Service. These divisions are listed separately in this directory.

**U.S. Department of Energy**
1000 Independence Avenue, SW
Washington, DC 20585
(800) DIAL–DOE
www.doe.gov
Contact: Public Information Officer

The DOE deals with both domestic and international issues relating to science and technology, energy efficiency, and the environment. The agency hosts numerous programs focused on the oil and natural gas industry, including the National Petroleum Council and Office of Fossil Energy.

**U.S. Department of the Interior**
1849 C Street, NW
Washington, DC 20240
(202) 208–3100
www.doi.gov
Contact: Secretary

The department is responsible for the administration of eight bureaus and offices relating to natural resources: the Bureau of Indian Affairs, Bureau of Land Management, Bureau of Reclamation, Minerals Management Service, National Park Service, Office of Surface Mining, U.S. Geological Survey, and U.S. Fish and Wildlife Service.

**U.S. Environmental Protection Agency**
1200 Pennsylvania Avenue, NW
Washington, DC 20460
(292) 272–0167
www.epa.gov
Contact: Administrator

EPA has ten regional offices that provide technical assistance, research and development, monitoring and enforcement, and implementation of federal legislation and regulations relating to air and water pollution, toxic and hazardous waste, indoor pollution, and other environmental concerns.

# UN Organizations

**UN Department for Economic and Social Affairs**
Commission on Sustainable Development
Two United Nations Plaza, Room DC2–2220
New York, NY 10017
(212) 963–8102
www.un.org/esa/sustdev
Contact: Secretary

Established by the UN General Assembly in December 1992, following the UN Conference on Environment and Development, the Commission on Sustainable Development reviews the implementation of international agreements, provides policy guidance, and supports innovative strategies for sustainability through two-year cycles focusing on specific themes.

**UN Environment Programme**
PO Box 67578
Nairobi, Kenya 00200
www.unep.org
Contact: Secretariat Executive Director

The UNEP is the only major UN agency headquartered in a Third World nation as part of an effort to globalize the importance of environmental protection. It provides leadership and encourages partnerships in caring for the environment through scientific advisory groups, a committee of permanent representatives, regional offices, and field programs.

**UN Food and Agriculture Organization**
Viale delle Terme di Caracalla
00100 Rome, Italy
www.fao.org/UNFAO
Contact: Secretariat

Since its founding in 1945, the FAO has led international efforts to defeat hunger, serving both developed and developing nations. It is a neutral forum for negotiation and policy debate. One of the FAO's major initiatives is to improve forestry practices by putting information within reach, sharing policy expertise, providing a meeting place for nations, and bringing knowledge to the field.

# 8

# Selected Print and Nonprint Sources

## Books

Adams, Jonathan. 2002. *The Future of the Wild: Radical Conservation for a Crowded World.* Boston: Beacon. 296 pp.

North America's wilderness is under attack, and the author uses the stories behind the Northern spotted owl, grizzly bear, mountain lion, and other species to illustrate how effective conservation strategies can provide wildlife corridors to connect the larger landscape.

Alverson, William S., Walter Kuhlmann, and David M. Waller. 1994. *Wild Forests: Conservation Biology and Public Policy.* Washington, DC: Island. 300 pp.

Forest conservation goes far beyond the protection and sustainability of a single natural resource. Management includes forest ecology, the conservation of biological diversity, and agency performance measures.

Anderson, Anthony B., and Clinton N. Jenkins. 2006. *Applying Nature's Design: Corridors as a Strategy for Biodiversity Conservation.* Irvington, NY: Columbia University Press. 256 pp.

**247**

Biological corridors, areas of land set aside to facilitate the movement of species, are gaining attention from ecologists as a way of maintaining landscape connectivity. Through the use of case studies in Costa Rica, Florida, the Canadian Rockies, Malaysia, and other examples, the authors identify the political and socioeconomic issues that must be considered.

Anderson, Terry L., and Donald R. Leal. 2001. *Free Market Environmentalism.* New York: Palgrave. 241 pp.

The author looks at a wide spectrum of potential solutions to natural resource issues, from reducing red tape and bureaucracy that tie up production, to marketing the by-products of waste management, to taking a free market approach to global conflicts.

Baden, John A., and Donald Snow, eds. 1997. *The Next West: Public Lands, Community, and Economy in the American West.* Washington, DC: Island. 272 pp.

Nearly a dozen leading writers offer an insightful vision of the future of the American West in a compilation of their views on what has gone wrong in the region, highlighting various methods of environmental stewardship.

Barlow, Maude. 1999. *Blue Gold: The Global Water Crisis and the Commodification of the World's Water Supply.* Sausalito, CA: International Forum on Globalization. 50 pp.

The term "blue gold" refers to the value of water as a natural resource, and this book explores the ways in which freshwater is being exploited through privatization. Water has become a commodity that can be sold and traded, leaving local populations out of the decision-making process altogether.

Bock, Carl E., and Jane H. Bock. 2000. *The View from Bald Hill: Thirty Years in an Arizona Grassland.* Berkeley: University of California Press. 197 pp.

One of the best ways to understand the issues involved in rangeland management is to understand the role of grassland ecology, and this book explores what happens when grazing is banned from an 8,000-acre plot that is restored naturally.

Bonnicksen, Thomas M. 2000. *America's Ancient Forests: From the Ice Age to the Age of Discovery.* New York: Wiley. 594 pp.

In order to understand the dynamics of forest controversies, it is useful to have the historical context provided in this book, which begins with the ice age forests and continues on through periods when forests were managed by indigenous peoples, the use of fire, Spanish explorers, forests of the colonies, and the role of trappers.

Bowles, Ian A., and Glenn T. Prickett, eds. 2001. *Footprints in the Jungle: Natural Resource Industries, Infrastructure, and Biodiversity Conservation.* New York: Oxford University Press. 332 pp.

By exploring how growing demands of resource extractive industries have affected conservation and development in the tropics, the authors presented in this volume review how oil and gas development, international trade, expanding markets, and minerals extraction have changed the world's view toward corporate responsibility.

Brick, Phillip D., and R. McGreggor Cawley, eds. 1996. *A Wolf in the Garden: The Land Rights Movement and the New Environmental Debate.* Lanham, MD: Rowman and Littlefield. 323 pp.

This balanced and objective book explores how government policy has changed with the advent of the wise use movement, county supremacy groups, and property rights organizations. It covers the major conflicts and the strategies being used by various stakeholders to influence the outcome of conflicts over national parks, old growth forests, and rangelands.

Brick, Phillip D., Donald Snow, and Sarah Van de Wetering, eds. 2001. *Across the Great Divide: Explorations in Collaborative Conservation and the American West.* Washington, DC: Island. 286 pp.

This edited volume represents the work of some of the country's preeminent scholars on environmental history and collaborative conservation, exploring how collaboration has influenced ranching interests and the culture of the West, and uses case studies from the well-known Quincy Library Group and Applegate Partnership to illustrate disputes over grizzly bear reintroduction and open space planning.

Browder, John O., and Brian J. Godfrey, eds. 1997. *Rainforest Cities: Urbanization, Development, and Globalization of the Brazilian Amazon*. New York: Columbia University Press. 424 pp.

Since the 1970s, the Brazilian Amazon has undergone significant changes and has been transformed into a frontier of new cities and regional development. The editors view the tropical rain forest from a unique perspective, exploring what they term "disarticulated urbanization" and its impact on the ecosystem.

Brunner, Ronald D., et al. 2005. *Adaptive Governance: Integrating Science, Policy, and Decision Making*. Irvington, NY: Columbia University Press. 368 pp.

Using five case studies from the West, this book examines different ways of making decisions about natural resource planning. Adaptive governance, unlike scientific management, emphasizes the need for open decision-making, the recognition of multiple interests, community-based initiatives, and an integrative science in addition to traditional science.

Brunner, Ronald D., et al. 2002. *Finding Common Ground: Governance and Natural Resources in the American West*. New Haven, CT: Yale University Press. 303 pp.

Using five case studies, the authors examine water management in Colorado, wolf recovery in the northern Rockies, bison management in Yellowstone, fisheries disputes, and forest policy in California.

Bryner, Gary C. 1998. *U.S. Land and Natural Resources Policy: A Public Issues Handbook*. Westport, CT: Greenwood. 292 pp.

This overview of the history and development of natural resource policies looks at the political factors that have shaped public land issues and how resources are valued.

Buckles, Daniel, ed. 2000. *Cultivating Peace: Conflict and Collaboration in Natural Resource Management*. Ottawa: International Development Research Center. 300 pp.

The competition for natural resources is a global one, as case studies from Africa, Asia, and Latin America illustrate, and new strategies, such as community-based natural resource management, are essential for a movement that emphasizes collaboration rather than conflict.

Bunnell, Fred L., and Jacklyn F. Johnson, eds. 1998. *Policy and Practices for Biodiversity in Managed Forests: The Living Dance.* Vancouver: University of British Columbia Press. 162 pp.

This edited volume uses the metaphor of the forest as a constantly, rapidly changing living dance to describe the complex ecological processes, changing values, and forces of time that affect life. At both the landscape level and in managed forests, biodiversity is affected by modification and adjustments from the scientific as well as political worlds.

Campari, João S. 2005. *The Economics of Deforestation in the Amazon: Dispelling the Myths.* Northampton, MA: Edward Elgar. 242 pp.

Within Brazil's Amazon River region, deforestation has been massive and recent, resulting in some of the most substantial environmental degradation in the world. While environmentally controversial, logging has also been criticized for the economic effect on the region's indigenous peoples, who depend on these resources for numerous uses.

Campbell, Greg. 2002. *Blood Diamonds: Tracing the Deadly Path of the World's Most Precious Stones.* Boulder, CO: Westview. 288 pp.

A journalist tells the story of the dark side of the diamond trade in Sierra Leone, where illegal smuggling and civil war have gone on virtually unnoticed by most of the rest of the world.

Carle, David. 2003. *Water and the California Dream: Choices for a New Millennium.* Berkeley: University of California Press. 255 pp.

One of the major conflicts in the West is the question of where water will come from in the future. This book explores this defining element of history, and the pressures of limitless growth on a limited source of supply.

Caviglia, Jill L. 1999. *Sustainable Agriculture in Brazil: Economic Development and Deforestation.* Northampton, MA: Edward Elgar. 160 pp.

Using case studies within the Ouro Preto region of Brazil, the author looks at economic issues rather than scientific ones, noting that market failure and the adoption of technology have competed with sustainable agriculture as reasons for the loss of tropical forest.

Chew, Sing C. 2001. *World Ecological Degradation: Accumulation, Urbanization, and Deforestation, 3000 B.C.–A.D. 2000.* Walnut Creek, CA: AltaMira. 217 pp.

The nine chapters in this book provide a comprehensive survey of historical periods and how the environment has been damaged over time. Starting with the Third Millennium and the Bronze Age, the author traces human impact in early Greece through the rise of the city-states and the Roman Empire to the growth of the Eastern Empire, the emerging economies of Europe, and the development of environmental consciousness.

Clarke, Jeanne Nienaber, and Daniel McCool. 1985. *Staking Out the Terrain: Power Differentials among Natural Resource Management Agencies.* Albany, NY: SUNY Press. 189 pp.

Although this book is best used for historical analysis, the authors provide a comprehensive background on the power of natural resource agencies, how expertise and the control of information affect policy, and the role of political and constituency support.

Cleuren, Herwig. 2001. *Paving the Road for Forest Destruction: Key Actors and Driving Forces of Tropical Deforestation in Brazil, Ecuador, and Cameroon.* Leiden, The Netherlands: Research School of Asian, African, and Amerindian Studies. 261 pp.

The author looks at three case studies of deforestation, identifying how stakeholders in each nation have increased the opportunities for commercial logging and illegal timber trade. The focus is on economic issues rather than the scientific reasons for changes in forest policy.

Collier, Paul, and Ian Bannon, eds. 2004. *Natural Resources and Violent Conflict: Options and Actions.* Washington, DC: World Bank. 428 pp.

This collection of essays seeks to explain the relationship between countries that regularly experience conflict and those whose economies are heavily dependent upon natural resources, using measures such as population, rebel financing, corruption, and resource reporting procedures.

Dagget, Dan, and Jay Dusard. 1995. *Beyond the Rangeland Conflict: Toward a West that Works.* Layton, UT: Gibbs Smith. 104 pp.

By arguing that the West's grazing conflicts must begin with a change in attitude among ranchers, environmental group members, and government range managers, the authors contend that the time, money, and effort being spent on arguments could be better used on more collaborative discussions.

Dana, Samuel T., and Sally K. Fairfax. 1980. *Forest and Range Policy: Its Development in the United States.* 2d ed. New York: McGraw-Hill. 458 pp.

This is one of the classic works in natural resources, covering both forests and grazing lands and the historical development of governmental agencies and regulations.

Daniels, Steven E., and Gregg B. Walker. 2001. *Working through Environmental Conflict: The Collaborative Learning Approach.* Westport, CT: Praeger. 299 pp.

The basis for most collaborative approaches to conflict management is public participation, and the authors show how, when combined with techniques like team building, citizen involvement can be used to deal with environmental problems at the local level.

Darley, Julian. 2005. *High Noon for Natural Gas: The New Energy Crisis.* White River Junction, VT: Chelsea Green. 266 pp.

This hard-hitting look at natural gas as an energy source outlines the implications of increased dependence and why it has the potential to cause serious environmental, political, and economic consequences.

d'Estrée, Tamara Pearson, and Bonnie B. G. Colby. 2004. *Braving the Currents: Evaluating Environmental Conflict Resolution in the River Basins of the American West.* New York, NY: Springer. 424 pp.

Using twenty-eight "success criteria," this is a systematic assessment of the elements that are needed for successful environmental conflict resolution, using disputes that have been resolved as case studies.

Devall, Bill, ed. 1995. *Clearcut: The Tragedy of Industrial Forestry.* San Francisco: Sierra Club Books. 291 pp.

The startling photographs that accompany the essays in this volume provide ample evidence of the visual impact of clear-cutting in both the United States and Canada, while authors call for a recognition of the intrinsic value of forests as well as their commodity value.

DeVilliers, Marq. 2001. *Water: The Fate of Our Most Precious Resource.* Boston: Mariner. 368 pp.

Although the author's tone is optimistic, the book provides an international perspective on both water quality and water scarcity. He calls for most of the usual changes to resolve the problems, from conservation and technological innovation to international cooperation and the reduction of wastefulness.

DeVoto, Bernard A. 2005. *DeVoto's West: History, Conservation, and the Public Good.* Athens: Ohio University Press. 275 pp.

A collection of essays that originally were published in *Harper's* from one of the nation's preeminent social commentators and advocates for the stewardship of American public lands. DeVoto was one of the most prolific and respected writers who commented on the evolution and change of the American West.

Dilsaver, Lary M., ed. 1994. *America's National Park System: The Critical Documents.* Lanham, MD: Rowman and Littlefield. 470 pp.

This book provides a compilation of the key documents that have punctuated the development of the national parks from the early beginnings in 1864 through the policies of the current system. It explores how various reports, proclamations, legislation, and executive orders have shaped the management and expansion of the parks over time.

Dobkowski, Michael N., and Isidor Wallimann, eds. 2002. *On the Edge of Scarcity: Environment, Resources, Population, Sustainability, and Conflict.* 2d ed. Syracuse, NY: Syracuse University Press. 296 pp.

From an economics perspective, there is a business application to natural resource conflicts that can explain problems such as sustainability and supply, the authors in the edited volume contend. Each essay explores these issues from a perspective that includes various geographic regions and problems.

Dolatyer, Mostafa, and Tim S. Gray. 2000. *Water Politics in the Middle East: A Context for Conflict or Cooperation?* New York: Palgrave Macmillan. 271 pp.

A controversial view of Middle Eastern water conflicts in which the authors argue that water scarcity is too important an issue to lead to warfare among competing water users. Still, there is little promise for the resolution of disputes that have divided the region for generations.

Dombeck, Michael P., Christopher A. Wood, and Jack E. Williams. 2003. *From Conquest to Conservation: Our Public Lands Legacy.* Washington, DC: Island. 232 pp.

The authors identify the key natural resource issues that have shaped U.S. environmental policy, including timber harvesting, water scarcity, grazing on public lands, and the efforts that have been made to resolve long-term conflicts.

Donahue, Debra L. 1999. *The Western Range Revisited: Removing Livestock from Public Lands to Conserve Native Biodiversity.* Norman: University of Oklahoma Press. 352 pp.

By examining scientific evidence of the aridity of the West and the ecological impact of cattle grazing, this book argues that there is an advantage to eliminating grazing that holds greater potential for benefiting biodiversity than any other single land use measure.

Elhance, Arun P. 1999. *Hydropolitics in the Third World: Conflict and Cooperation in International River Basins.* Washington, DC: U.S. Institute of Peace Press. 309 pp.

Shared river basins constitute nearly 50 percent of the world's landmass, leading to intense competition over water resources. Case studies of six river basins explain the reasons behind water conflict in the developing world.

Elmusa, Sharif S. 1998. *Water Conflict: Economics, Law and Palestinian-Israeli Water Resources.* Beirut, Lebanon: Institute for Palestine Studies. 408 pp.

This neutral and objective view of the major water conflict in the Middle East explains the policy issues from the perspective of both parties, outlining how each side supports its arguments for dealing with the control of the water supply.

Franke, Mary Ann. 2005. *To Save the Wild Bison: Life on the Edge in Yellowstone.* Norman: University of Oklahoma Press. 368 pp.

Changing attitudes about wildlife, and their place in protected areas, are examined in this book, which examines the ecological and political aspects of the controversy over bison management policies.

Gale, Fred P. 1998. *The Tropical Timber Trade Regime.* New York: St. Martin's. 287 pp.

The author begins with an overview of the tropical rain forest crisis, and then explains how international regimes, including the International Tropical Timber Agreement of 1983 and subsequent conflicts over its enforcement, have led to forest degradation.

Gausset, Quentin, Michael A. Whyte, and Torben Birch-Thomsen, eds. 2005. *Beyond Territory and Scarcity: Exploring Conflicts over Natural Resource Management.* Uppsala, Sweden: Nordic Africa Institute. 218 pp.

Written by geographers and anthropologists, this collection of essays deals with the attainment of sustainable natural resource management in sub-Saharan Africa, examining issues in the context of rapid population growth within the region.

Gedicks, Al. 2001. *Resource Rebels: Native Challenges to Mining and Oil Corporations.* Cambridge, MA: South End. 241 pp.

This review of how indigenous peoples have attempted to deal with multinational corporations seeking to develop mineral and oil resources also explores the successes and failures of native peoples facing legal and financial giants.

Gorte, Ross W., ed. 2003. *National Forests: Current Issues and Perspectives.* Hauppauge, NY: Nova Science. 181 pp.

The author questions how the United States can develop a balance between the protection of timberland in the national forests and the growing problems of diminishing federal resources, the need for a steady timber supply, the legacy of a massive system of forest roads, and legislative mandates for stewardship contracting.

Grafton, R. Quentin, et al. 2004. *The Economics of the Environment and Natural Resources*. Malden, MA: Blackwell. 503 pp.

Using models, systems, and dynamics, the authors examine the bioeconomics of natural resources, including fisheries, forests, water, and the environmental valuation of nonrenewable resources.

Grusin, Richard A. 2004. *Culture, Technology, and the Creation of America's National Parks*. Cambridge: Cambridge University Press. 232 pp.

The author investigates how the establishment of the national parks helped lead to an American national identity after the Civil War.

Halvorson, William L., and Gary E. Davis, eds. 1996. *Science and Ecosystem Management in the National Parks*. Tucson: University of Arizona Press. 362 pp.

The chapters in this edited volume approach natural resources from the perspective of conservation and ecosystem management, identifying how government policies have often conflicted with science.

Hart, Matthew. 2003. *Diamond: A Journey to the Heart of an Obsession*. New York: Walker and Company. 288 pp.

The author traces the diamond frenzy that struck Canada in the 1990s, with a background on the history of the South African diamond cartel, the role of Brazilian dealers, polishing rooms in New York, and sorting rooms in London.

Hartzog, George B., Jr. 1988. *Battling for the National Parks*. Mt. Kisco, NY: Moyer Bell. 284 pp.

For those looking for a comprehensive history of both the National Park Service and the development of parks and reserves, this book provides both excellent biographical information and objective analysis.

Harvey, Mark. 2005. *Wilderness Forever: Howard Zahniser and the Path to the Wilderness Act.* Seattle: University of Washington Press. 328 pp.

If one person were said to be responsible for the passage of the 1964 Wilderness Act, it would undoubtedly be Howard Zahniser, an outdoorsman who pulled together the fragmented factions of the early American environment movement to lobby tirelessly for a law that is the foundation for contemporary wilderness preservation.

Hays, Samuel P. 1959. *Conservation and the Gospel of Efficiency: The Progressive Conservation Movement 1890–1920.* Cambridge, MA: Harvard University Press. 297 pp.

Hays, one of the country's most noted environmental historians, focuses on Gifford Pinchot, the turn of the twentieth century, and the utilitarian view of natural resources, contrasting those views with the reforms brought on by the progressive movement, preservationists like John Muir, and the scientific management revolution.

Heinberg, Richard. 2003. *The Party's Over: Oil, War, and the Fate of Industrial Societies.* Gabriola Island, BC: New Society. 288 pp.

Heinberg predicts chaos unless the United States is willing to join other countries to implement a global program of resource conservation and sharing, and explains what it will take to reduce the world's reliance upon fossil fuels.

Holechek, Jerry L., et al. 2003. *Natural Resources: Ecology, Economics, and Policy.* 2d ed. Upper Saddle River, NJ: Prentice Hall. 761 pp.

This is a detailed history of natural resource management in the United States, natural resources, and international development. The authors advocate an integrated approach from an economic perspective.

Holechek, Jerry L., Rex D. Pieper, and Carlton H. Herbel. 2003.

*Range Management: Principles and Practices.* 5th ed. Upper Saddle River, NJ: Prentice Hall. 624 pp.

The science of range management includes a variety of topics, including the ecology of plants and wildlife, choosing an appropriate form of grazing, the characteristics of rangeland, inventory and monitoring, and livestock production, each of which is covered in this introductory book.

Holmes, Thomas P., et al. 2000. *Financial Costs and Benefits of Reduced Impact Logging relative to Conventional Logging in the Eastern Amazon.* Alexandria, VA: Tropical Forest Foundation. 48 pp.

This short report, produced under the sponsorship of a nongovernmental organization, proposes innovative techniques that can be used as an alternative to traditional logging methods while still returning a profit for local economies.

Hunt, Constance E. 2004. *Thirsty Planet: Strategies for Sustainable Water Management.* New York: Palgrave Macmillan. 256 pp.

The author suggests that nature is the source of water and only by making the conservation of nature an absolute priority will we have water in the future for human use. The book looks at the complexity of the problem, providing an array of ideas, information, case studies, and ecological information.

Jackson, Donald C. 2005. *Building the Ultimate Dam: John S. Eastwood and the Control of Water in the West.* Norman: University of Oklahoma Press. 352 pp.

Much of the water policy of the West was built around the idea that "bigger is better," and the author outlines how changes in engineering and attitudes led to new policies and designs.

Jeffery, Roger, and Bhaskar Vira, eds. 2002. *Conflict and Cooperation in Participatory Natural Resource Management.* New York: Palgrave Macmillan. 264 pp.

Historically, natural resources have been managed by centralized government institutions with the belief that this increases equity and efficiency. This volume questions whether those goals can be met at lower cost and more democratically through the use of collaborative management agreements.

Jensen, Derrick, and George Draffan. 2003. *Strangely Like War: The Global Assault on Forests.* White River Junction, VT: Chelsea Green. 185 pp.

With a foreword by environmental leader Vandana Shiva, this book provides both a historical account of deforestation and an analysis of the environmental impact of clear-cutting on forest ecology.

Johnson, Elizabeth A., and Michael W. Klemens, eds. 2005. *Nature in Fragments: The Legacy of Sprawl.* Irvington, NY: Columbia University Press. 400 pp.

While sprawl is usually considered in an urban context, the essays in this book focus on the role of development and biological diversity. Topics include land use and freshwater wetlands, bees and pollination, disease, wide-ranging species, and the role of public awareness.

Just, Richard E., and Sinaia Netanyahu, eds. 1998. *Conflict and Cooperation on Trans-Boundary Water Resources.* Boston: Kluwer Academic. 432 pp.

The book examines the nature of intrastate, interstate, and international disputes over water quality and quantity from a global perspective, noting that transboundary water resources are often a cause of conflict worldwide.

Kamieniecki, Sheldon. 2006. *Corporate America and Environmental Policy: How Often Does Business Get Its Way?* Palo Alto, CA: Stanford University Press. 348 pp.

The six cases in the book outline the way in which private companies involved in disputes over natural resource management were able to influence environmental policy in all three branches of government. A conceptual framework is offered, as well as arguments about the ability of business interests to influence policy-making.

Klare, Michael T. 2001. *Resource Wars: The New Landscape of Global Conflict.* New York: Metropolitan. 289 pp.

The author contends that in the coming decades, wars will be fought over increasingly scarce natural resources, especially over oil and water. The book also explores conflicts over timber, gems, and minerals from an international perspective.

Klyza, Christopher McGrory. 1996. *Who Controls Public Lands?: Mining, Forestry, and Grazing Policies, 1870–1990*. Chapel Hill: University of North Carolina Press. 211 pp.

Klyza's history of policy in these three areas provides an overview of the context in which laws have been enacted, social and cultural paradigms have been changed, and entire industries have changed through governmental regulation.

Koontz, Tomas M. 2002. *Federalism in the Forest: National versus State Natural Resource Policy.* Washington, DC: Georgetown University Press. 232 pp.

The strengths and weaknesses of forest policy in the United States balance between federal agency control and management by state government. Comparing four forest pairs, the book examines revenue sharing, public participation, and environmental protection as measures of policy performance differences.

Larmer, Paul, ed. 2003. *Give and Take: How the Clinton Administration's Public Lands Offensive Transformed the American West.* Paonia, CO: High Country News. 230 pp.

Focused on the eight years of public lands policy under President Bill Clinton, this book highlights the issues of rural land use, national parks and reserves, and the conservation of natural resources.

Lee, Robert G., and Donald R. Field, eds. 2005. *Communities and Forests: Where People Meet the Land.* Corvallis: Oregon State University Press. 320 pp.

As forest science changes, so too do the values that influence forest policy-making. The authors identify four major types of forestry: managing solely for wood, managing for benefits ranging from watersheds to food crops, protecting natural forests, and urbanizing forests.

Lewicki, Roy J., Barbara Gray, and Michael Elliott, eds. 2003. *Making Sense of Intractable Environmental Conflicts: Frames and Cases.* Washington, DC: Island. 469 pp.

By examining case studies dealing with natural resources, water, toxics, and growth management, the authors show how disputes can be interpreted or framed to explain what the conflict is about, who is involved, the motivations of the parties, and how it should be solved.

Lindholdt, Paul, and Derrick Knowles, eds. 2005. *Holding Common Ground: The Individual and Public Lands in the American West.* Spokane: Eastern Washington University Press. 152 pp.

This is an anthology of thirty Western writers who provide an insight into their perspectives on public lands, ranging from Carolyn Kremer's essay on civil disobedience in the Arctic National Wildlife Refuge to Bruce Eilerts's "War among the Saguaros."

Lowenthal, David. 2002. *George Perkins Marsh: Prophet of Conservation.* Seattle: University of Washington Press. 632 pp.

Considered one of the first to recognize the dangers of human degradation of the environment, and possible reforms, Marsh sounded the clarion call for change in 1864. This biography identifies Marsh's broad scholarship, his political and diplomatic career, and his ominous warnings about the future if humans fail to exercise stewardship over the earth.

Mateo, Rony, ed. 2004. *American National Parks: Current Issues and Developments.* Hauppauge, NY: Novinka. 122 pp.

While the original goal of the U.S. national park system may have been to preserve scenic beauty and provide opportunities for learning, solace, and recreation, today's protected areas, administered by the National Park Service, face numerous challenges. The authors of this edited volume explore how national parks deal with reduced funding, a crumbling infrastructure, and issues such as mining in the parks.

Mayer, Kenneth R. 2001. *With the Stroke of a Pen: Executive Orders and Presidential Power.* Princeton, NJ: Princeton University Press. 293 pp.

Very few researchers have examined the role of the executive order in U.S. history, and this book is especially timely in explaining how President Bill Clinton, among others, used his powers to establish new environmental policies.

McKinney, Matthew, and William Harmon. 2004. *The Western Confluence: A Guide to Governing Natural Resources*. Washington, DC: Island. 297 pp.

This contemporary overview of water disputes in the West looks at the historical principles of water rights, the role of citizen groups, and the need for dialog among stakeholders such as snowmobile users, the U.S. Forest Service, environmental organizations, and the Environmental Protection Agency.

Mittermeier, Russell A., et al. 2005. *Transboundary Conservation: A New Vision for Protected Areas*. Washington, DC: Conservation International. 372 pp.

The authors of this book note that there are now 818 protected regions in 112 countries that make up 188 transboundary areas that straddle international boundaries. Focusing on 28 of these areas spanning the globe (including Antarctica), they point out the potential for new types of management that may reduce international conflict while protecting biodiversity.

Moran, Emilio F., and Elinor Ostrom, eds. 2005. *Seeing the Forest and the Trees: Human-Environment Interactions in Forest Ecosystems*. Cambridge, MA: MIT Press. 442 pp.

Using an interdisciplinary approach, the editors have selected more than a dozen themes relating to the ways in which data on deforestation is collected, the processes of forest change, land-cover change information, the use of geographic information systems, and cross-continental comparisons.

Mutz, Kathryn M., Gary C. Bryner, and Douglas S. Kenney, eds. 2002. *Justice and Natural Resources: Concepts, Strategies and Applications*. Washington, DC: Island. 368 pp.

By using the theme of environmental justice, the editors look at conflicts over water, mineral development, and tribal sovereignty. They expand on the research relating to environmentally hazardous facilities to provide a more expansive view of natural resource decisions.

Nielsen, John. 2005. *Condor: To the Brink and Back—The Life and Times of One Giant Bird*. New York: HarperCollins. 257 pp.

The California condor represents, symbolically and practically, one of the most important examples of humanity's efforts to control nature. From the time when the last wild condor was captured in 1987 to now, when there are more than 200 birds living in the wild, the species has been mired in controversy over whether it was better to allow the bird to go extinct, or manipulate nature by keeping some in zoos and preserves in the hope of helping it to recover.

O'Leary, Rosemary, and Lisa Bingham, eds. 2003. *The Promise and Performance of Environmental Conflict Resolution.* Washington, DC: Resources for the Future Press. 400 pp.

This book provides empirical research along with insights from some of the most experienced practitioners in environmental conflict resolution, beginning with concepts and methods used by scholars in political science, public administration, regional planning, psychology, anthropology, and law.

Owen, Oliver, Daniel D. Chiras, and John P. Reganold. 1998. *Natural Resource Conservation: Management for a Sustainable Future.* Upper Saddle River, NJ: Prentice Hall. 594 pp.

While many of the books dealing with natural resources approach the topic from the perspective of conflict, these authors use sustainability as the measurement of success and failure.

Palm, Cheryl L., et al., eds. 2006. *Slash-and-Burn Agriculture: The Search for Alternatives.* Irvington, NY: Columbia University Press. 480 pp.

Estimating that more than 80,000 square miles of tropical rain forest are destroyed each year, a consortium of international institutions attempts to address the problems of conducting this form of agriculture in sensitive areas. Site-specific examples are included for the Brazilian Amazon, Cameroon, Indonesia, Peru, and Thailand.

Palo, Matti, and Heidi Vanhanen, eds. 2000. *World Forests from Deforestation to Transition?* Boston: Kluwer Academic. 216 pp.

This global survey of deforestation covers forest transition in various regions, from China and New England to the Brazilian Amazon and Indonesia. The authors cover both the scientific perspective and the sociopolitical view, including agricultural expansion and cultural histories.

Pearce, Fred. 2005. *When the Rivers Run Dry: Water, the Defining Crisis of the Twenty-First Century.* Boston: Beacon. 336 pp.

The author analyzes the prediction that, by 2025, water scarcity will cut global food production by more than the current U.S. grain harvest. Pearce traveled to more than thirty countries to research this book, and he identifies the key rivers that are part of the water crisis. He believes that the picture is dire, but not without solutions.

Pinchot, Gifford. 1937. *The Training of a Forester.* Philadelphia: J.B. Lippencott. 129 pp.

America's first chief of the U.S. Forest Service shares his views on the need for reform.

Posey, Darrell A., and Michael J. Balick, eds. 2006. *Human Impacts on Amazonia: The Role of Traditional Ecological Knowledge in Conservation and Development.* Irvington, NY: Columbia University Press. 368 pp.

This series of essays represents the work of environmental scientists, anthropologists, and botanists who document the role of human development in this sensitive ecosystem. Local knowledge and values are seen as one of the best ways to learn how to protect the Amazon and avoid further degradation of the tropical forest.

Postel, Sandra. 1999. *Pillar of Sand: Can the Irrigation Miracle Last?* New York: W. W. Norton and Co. 313 pp.

Calling water scarcity now the single most serious threat to global food production, Postel, director of the Global Water Policy Project, shows how the combination of an increase in population to 8 billion by 2030, coupled with the fact that the earth's freshwater supply is not where that population is living, demands new ways of allocating water, such as making irrigation more efficient.

Postel, Sandra. 1997. *Last Oasis: Facing Water Scarcity.* New York: W. W. Norton. 191 pp.

The editors challenge the commonly held notion that water is in endless supply, arguing that we have entered an era of water scarcity. Shortages will impact economics, the world's ecology, and politics, but there are many technological solutions that can reduce conflict over water.

Postel, Sandra, and Brian Richter. 2003. *Rivers for Life: Managing Water for People and Nature.* Washington, DC: Island. 253 pp.

This global perspective on the challenges of water management focuses on the disruption of rivers by diverting water through 45,000 large dams worldwide, altering the natural flow of hydrologic cycles, and destroying wildlife habitat.

Pynn, Larry. 2000. *Last Stands: A Journey through North America's Vanishing Ancient Rainforests.* Corvallis: Oregon State University Press. 212 pp.

The Northwest coast of North America has a different type of old growth forest than those traditionally associated with the Pacific Northwest. This book deals primarily with the ecology of ancient rain forests but also explains their role in forest conservation.

Renner, Michael. 2002. *The Anatomy of Resource Wars.* Washington, DC: Worldwatch Institute. 91 pp.

This book explores how natural resource abundance may fuel violent conflict, whether to control resources (such as diamonds or oil), and how those conflicts are being used by corrupt regimes to bankroll further violence at the expense of the environment.

Righter, Robert W. 2005. *The Battle over Hetch Hetchy: America's Most Controversial Dam and the Birth of Modern Environmentalism.* New York: Oxford University Press. 328 pp.

Considered by many to be the first major environmental battle of the twentieth century, the flooding of California's Hetch Hetchy Valley to provide drinking water for San Francisco epitomizes the struggle between conservationists and preservationists. This book is unique because it includes coverage from the completion of the dam in 1934 to the 1998 movement to restore the valley.

Roberts, Janine. 2004. *Glitter and Greed: The Secret World of the Diamond Cartel.* New York: Disinformation Company. 352 pp.

Roberts, an Australian investigative journalist, provides an exposé of the De Beers diamond operations, the diamond industry, and the conflicts created by diamond traders and miners with aboriginal people.

Robinson, John, and Elizabeth Bennett, eds. 1999. *Hunting for Sustainability in Tropical Forests.* New York: Columbia University Press. 1,000 pp.

This extensive volume presents the results of numerous studies conducted by biological and social scientists. The emphasis is on the role of hunting of wildlife and native peoples, challenges to resource management, and the economic issues associated with sustainability in a resource-rich area.

Roddick, Anita, and Brooke S. Biggs. 2005. *Troubled Water: Saints, Sinners, Truth and Lies about the Global Water Crisis.* White River Junction, VT: Chelsea Green. 138 pp.

A global perspective on the politics of water, including problems, solutions, and resources offered for the general reader as a way of becoming actively involved.

Sabatier, Paul, et al., eds. 2005. *Swimming Upstream: Collaborative Approaches to Watershed Management.* Cambridge, MA: MIT Press. 328 pp.

Historically, watershed management in the United States has been based on a traditional, top-down hierarchical management model. This book examines collaborative projects from a social, political, and economic perspective, and the role of negotiation in solving watershed problems.

Satterfield, Terre. 2003. *Anatomy of a Conflict: Identity, Knowledge, and Emotion in Old-Growth Forests.* Vancouver: University of British Columbia Press. 198 pp.

While the timber wars of the Pacific Northwest are usually described as a dispute between jobs and the environment, this ethnographic study shows that the debate is centered on cultural patterns and social movements.

Schroth, Gotz, ed. 2004. *Agroforestry and Biodiversity Conservation in Tropical Landscapes.* Washington, DC: Island. 523 pp.

This lengthy volume is composed of twenty chapters that cover the entire spectrum of tropical forest resources. The authors examine fragmented tropical landscapes, the ecological economics of agroforestry, specific crops such as cocoa and coffee, tools such as live fences and windbreaks, and biodiversity as both burden and natural capital.

Sellars, Richard W. 1997. *Preserving Nature in the National Parks.* New Haven, CT: Yale University Press. 380 pp.

This book analyzes how the National Park Service faces the dilemma of how and what to preserve—scenery, landscapes, mammals, trees and plants, or the entire ecosystem, and how the emphasis on recreational tourism has been reflected in the agency's organizational power structure.

Sherman, Martin. 1999. *The Politics of Water in the Middle East: An Israeli Perspective on the Hydro-Political Aspects of the Conflict.* New York: St. Martin's. 106 pp.

The author notes that while water is an enduring conflict in the Middle East, little attention has been paid to the controversies over which nations would control Israel's water supplies should the peace accords be finalized.

Simmons, Matthew R. 2005. *Twilight in the Desert: The Coming Saudi Oil Shock and the World Economy.* Hoboken, NJ: John Wiley and Sons. 448 pp.

This controversial analysis of Saudi oil supplies poses the question of how long the countries in the Middle East can meet world demand for oil, and how countries will adapt to oil scarcity. The research and statistics have been called into question by some reviewers, although many agree that the questions about supplies need to be asked.

Smith, Duane A. 1987. *Mining America: The Industry and the Environment, 1800–1980.* Lawrence: University Press of Kansas. 210 pp.

This book covers the mining industry's development from a number of perspectives, from the scars resulting from the nineteenth-century mining districts to the environmental impacts of contemporary extractive resource industries.

Sponsel, Leslie E., Thomas N. Headland, and Robert C. Bailey. 1996. *Tropical Deforestation: The Human Dimension.* New York: Columbia University Press. 365 pp.

Part of a series in conservation science, this book looks at the social aspects of forest ecology and management in tropical forests. While based on biological principles, the book covers various resource management issues.

Stegner, Wallace. 1974. *The Uneasy Chair: A Biography of Bernard DeVoto.* Garden City, NY: Doubleday. 464 pp.

One of the West's most respected environmental historians writes about a fellow historian and conservationist, "Benny" DeVoto, whose columns in *The Saturday Review of Literature* and *Harper's* became a platform for the acclaimed writer's views on management of public lands.

Sterner, Thomas. 2003. *Policy Instruments for Environmental and Natural Resource Management.* Washington, DC: Resources for the Future. 504 pp.

An examination of governmental policy for the conservation of natural resources, with an extensive list of bibliographical references.

Strohmeyer, John. 1993. *Extreme Conditions: Big Oil and the Transformation of Alaska.* New York: Simon and Schuster. 287 pp.

This analysis of the petroleum industry looks at the role of internal demand and international trade, identifying many of the environmental impacts of oil on the Alaskan people and their resources.

Struhsaker, Thomas T. 1997. *Ecology of an African Rain Forest: Logging in Kibale and the Conflict between Conservation and Exploitation.* Gainesville: University Press of Florida. 434 pp.

Most of the attention to tropical forests has centered on Latin and Central America, but this author uses the case study of the Kibale Forest Reserve in Uganda. The book covers the effects of logging on both trees and biodiversity, with concluding recommendations on tropical rain forest management policy and practice.

Susskind, Lawrence, and Jeffrey Cruikshank. 1987. *Breaking the Impasse: Consensual Approaches to Resolving Public Disputes.* New York: Basic. 276 pp.

One of the first major books on the emerging tools of conflict management, this volume explores the role of consensus as a way of dealing with some of the most difficult political and social situations and issues now being faced.

Sutter, Paul S. 2002. *Driven Wild: How the Fight against Automobiles Launched the Modern Wilderness Movement.* Seattle: University of Washington Press. 343 pp.

Sutter explains how the conflict to do as one wishes and the freedom to experience solitude developed in the United States. He notes the connection between our love for nature and our love for cars and how that led to a perception that cars were the enemy of wilderness, disassociating us from the peace of the outdoors.

Tal, Alon. 2006. *Speaking of Earth: Environmental Speeches that Moved the World.* New Brunswick, NJ: Rutgers University Press. 277 pp.

The author notes that the one thing that environmental activists around the world have in common is their voices. This collection of twenty speeches includes background and biographical information on Rachel Carson and David Brower in the United States and global leaders such as Mostafa Tolba and the Dalai Lama.

Vallentine, John F. 2001. *Grazing Management.* 2d ed. San Diego, CA: Academic. 560 pp.

Defined as the manipulation of animal grazing to achieve desired results based on plant, land, and economic responses, grazing management goes beyond simply maximizing the amount and quality of forage needed by livestock, as this somewhat technical manual explains.

Vaughn, Jacqueline, and Hanna J. Cortner. 2005. *George W. Bush's Healthy Forests: Reframing the Environmental Debate.* Boulder: University Press of Colorado. 231 pp.

The signing of the 2003 Healthy Forests Restoration Act is a case study in understanding how the political process can be used to move an administration's agenda forward. In this instance, President George W. Bush and his appointees used the legislative and regulatory processes, along with fortuitous natural events, to change the direction of policy and public participation in the decision-making process.

Ward, Diane Raines. 2003. *Water Wars: Drought, Flood, Folly, and the Politics of Thirst.* New York: Riverhead. 320 pp.

The author contends that the public does not understand the complicated nature of the world's water wars, whether the conflict is in the Indus Basin or the Arabian Sea. She examines the role of dams, levees, rivers, and irrigation systems to show how difficult it is to find solutions that match the unique aspects of each type of problem.

Wescoat, James L., and Gilbert F. White. 2003. *Water for Life: Water Management and Environmental Policy.* New York: Cambridge University Press. 322 pp.

This book explores the role of water in ecosystems, from groundwater, lakes, and wetlands to river channels and floodplains, explaining that water can be managed through a variety of governmental and nongovernmental structures and policies.

Whiteman, Michael R. 1994. *An Assessment of Management Strategies for Public Natural Resource Conflict.* Moscow: University of Idaho. 134 pp.

This Ph.D. thesis explores the possibilities and pitfalls of using environmental mediation as a way of resolving public and private conflicts over natural resources.

Wondolleck, Julia M., and Steven L. Yaffee. 2000. *Making Collaboration Work: Lessons from Innovation in Natural Resource Management.* Washington, DC: Island. 277 pp.

Using more than 200 case studies, the authors explain the role of collaboration in natural resource management conflicts, the barriers that must be understood and overcome, and eight themes that characterize successful efforts.

Wood, Charles H., and Roberto Porro, eds. 2002. *Deforestation and Land Use in the Amazon*. Gainesville: University Press of Florida. 385 pp.

The Amazon is one of the most studied regions in the world for understanding tropical deforestation, and the editors provide a unique perspective by identifying not only the loss of natural resources but also the ways in which land is subsequently used.

Wuerthner, George, ed. 2006. *Wildfire: A Century of Failed Forest Policy*. Washington, DC: Island. 340 pp.

The book views wildfire from the perspective of its ecological, economic, and social/political impact, with essays from twenty-five fire ecology experts who contend that the twentieth century's policies of fire suppression have resulted in fuel overload. The major recommendation is to return fire to the landscape so that it will continue to be a natural part of ecological processes.

Wuerthner, George, and Mollie Yoneko Matteson, eds. 2002. *Welfare Ranching: The Subsidized Destruction of the American West*. Washington, DC: Island. 346 pp.

This series of essays portrays the economic and ecological impact of livestock ranching on the 300 million acres of public lands, with extensive photographs of natural resource damage, erosion, and spreading invasive species.

Yaffee, Steven L. 1994. *The Wisdom of the Spotted Owl: Policy Lessons for a New Century*. Washington, DC: Island. 430 pp.

This classic case study of the spotted owl controversy provides an in-depth history of the conflict, the strategies used by stakeholders, and how various groups used litigation and other tools to produce a controversial outcome for all involved.

Young, Herbert C. 2003. *Understanding Water Rights and Conflicts*. Denver, CO: BurgYoung. 275 pp.

The law of water rights is sometimes difficult to comprehend, but it is key to understanding the complexities of legal precedents and rulings that have often led to conflict over water scarcity.

# Monographs, U.S. and International Government Publications

National Research Council. 2000. Committee on Environmental Issues in Pacific Northwest Forest Management, Board on Biology, Commission on Life Sciences. *Environmental Issues in Pacific Northwest Forest Management.* Washington, DC: National Academy Press. 259 pp.

This committee, primarily composed of research scientists, takes a different look at old growth forests in its examination of rural communities, sustainability, fire, and landscape dynamics. More valuable to some are the committee's conclusions and recommendations.

UN Educational, Scientific, and Cultural Organization. 2004. *Beyond Tropical Deforestation: From Tropical Deforestation to Forest Cover Dynamics and Forest Development.* Paris: UNESCO. 488 pp.

The editor of this volume has placed an emphasis on the role of international cooperation as a mechanism for reducing deforestation in tropical areas, examining both scientific and political methods to resolve problems.

U.S. Department of Agriculture, Committee of Scientists. 1999. *Sustaining the People's Lands: Recommendations for Stewardship of the National Forests and Grasslands into the Next Century.* Washington, DC: U.S. Department of Agriculture. 193 pp.

This widely heralded report became the basis for much of the federal government's forest policies for the twentieth century, even though it was considered controversial by stakeholders on all sides of the forest resource debate.

U.S. Department of Agriculture, U.S. Forest Service. 2005. *100 Years of Conservation: For the Greatest Good.* Washington, DC: U.S. Department of Agriculture. 9 pp.

Part of the centennial celebration of the U.S. Forest Service, this document outlines the development of forest policy from a historical perspective.

U.S. Department of the Interior. 2004. *Cooperative Conservation: Success through Partnerships.* Washington, DC: U.S. Department of the Interior. 45 pp.

Examines the role of citizen participation in the conservation of natural resources, and the role of communication and volunteers.

U.S. Government Accountability Office. 2005. *Wildland Fire Management: Progress and Future Challenges, Protecting Structures, and Improving Communications.* Washington, DC: U.S. Government Accountability Office. 23 pp.

The report recommends that although significant progress has been made in responding to wildland fires, federal agencies still need to develop a strategic plan that identifies options and the funding needed to deal with risks.

United States. Congress. House. Committee on Agriculture. Subcommittee on Forests, Family Farms, and Energy. 1989. *Management of Old-Growth Forests of the Pacific Northwest:* Joint Hearings before the Subcommittee on Forests, Family Farms, and Energy of the Committee on Agriculture and the Subcommittee on National Parks and Public Lands of the Committee on Interior and Insular Affairs, House of Representatives, 101st Cong., 1st Sess., June 20 and 22. 593 pp.

This lengthy volume of congressional testimony focuses on the impact of changes in old growth forest management on small communities, employment, and the economy of the Pacific Northwest.

United States. Congress. Senate. Committee on Environment and Public Works. Subcommittee on Environmental Protection. 1992. *Conservation of the Northern Spotted Owl:* Hearing before the Subcommittee on Environmental Protection of the Committee on Environment and Public Works, United States Senate, 102nd Cong., 2d. Sess., May 13. 157 pp.

This Senate hearing explains many of the key issues and identifies the major stakeholders in the spotted owl controversy.

# Articles

Albers, Heidi J., Amy W. Ando, and Daniel Kaffine. 2004. "Land Trusts in the United States: Analyzing Abundance." *Resources* (spring): 10–13.

Land trusts have emerged as a way of protecting tracts of valuable watersheds and wildlife habitat, open space in highly developed areas, and recreational use areas. The authors question whether this strategy is a sound mechanism for preserving land in addition to existing programs administered by agencies such as the National Park Service.

Barringer, Felicity. 2004. "Judge's Ruling on Yellowstone Keeps It Open to Snowmobiles." *New York Times*, October 16, p. A9.

Despite the Clinton administration's efforts to phase out snowmobiles in Yellowstone National Park, a federal judge in Wyoming ruled that the public had not been properly consulted and that the government had not followed proper procedures.

Biswas, Asit K. 2005. "An Assessment of Future Global Water Issues." *International Journal of Water Resources Development* 21, no. 2 (June): 229–237.

The author contends that the global emphasis on water scarcity is misdirected, even though some countries may have difficulty managing water supplies. Instead, the more serious problem is rapid water quality deterioration, which is not being addressed in resource debates.

Borod, Jed. 2005. "Fools Gold: Illegal Mining Linked to Conflict in the Democratic Republic of Congo." *International Enforcement Law Reporter* 21, no. 10 (October): 410.

Natural resource exploitation has been the focus of numerous nongovernmental organizations, with increased attention paid to the Democratic Republic of the Congo, where there have been more than 3 million deaths since 1998. Many of these deaths coincide with the development of militias protecting illegal gold mining and smuggling operations.

Breslau, Karen. 2004. "Working to Save the West." *Newsweek* (October 18): 56.

Groups like the Conservation Fund are working with ranchers to save endangered landscapes from builders through conservation easements and land purchases.

Campbell, Greg. 2002. "Blood Diamonds." *Amnesty Magazine* (fall): 4–7.

This story of a fifteen-year-old boy, a soldier in Sierra Leone's Revolutionary United Front, shows the underlying violence that is part of the attempts to control the trade in raw diamonds that has paralyzed the region since 1991.

Cockburn, Andrew. 2002. "Diamonds: The Real Story." *National Geographic* 201, no. 3 (March): 2–35.

The trail of diamonds from the ground to jewelry is not a simple one, as the author explains in this narrative about the complications of the diamond industry.

Davis, Charles. 1998. "Gold or Green? Efforts to Reform the Mining Law of 1872." *Natural Resources and Environmental Administration* 19 (February): 2–4.

Davis looks at one of the oldest natural resource statutes in the United States, and the ways in which the original plan to settle the West by granting mineral rights patents is no longer relevant. Despite more than 125 years of attempts to reach agreement on change, the original law is still intact.

Davis, Phillip. 1991. "Cry for Preservation, Recreation Changing Public Land Policy." *Congressional Quarterly Weekly Report* (August 3): 2151.

Although grazing policy was never a key element of the presidential campaign, the author argues that more candidates are taking a second look at priorities for the use of public land as a reaction to the Reagan administration's policies.

Feitelson, Eran. 2000. "The Ebb and Flow of Arab-Israeli Water Conflicts: Are Past Confrontations Likely to Resurface?" *Water Policy* 2, nos. 4–5: 343–363.

This article is a useful look back at the confrontations over the diversion of water in the Jordan River basin, which are often cited as resource-based conflicts. The author suggests that water may increasingly be a basis for confidence-building cooperation rather than confrontation.

Flora, Gloria. 2000. "Toward a Civil Discourse: The Need in Public Management." *Public Land and Resources Law Review* 21: 23–32.

One of the key elements of dealing with natural resource conflicts is the ability to communicate our views with civility and respect, the author contends, and failure to do so exacerbates conflict and shifts attention away from the real issues.

Garvin, Cosmo. 2004. "Old Growth Trees to Fall in the Sierra." *High Country News,* March 1, p. 4.

Among the most contentious issues relating to timber management is the protection of old growth timber, and nowhere is that conflict better represented than in efforts to log big trees in California.

Gleick, Peter H. 1993. "Water and Conflict: Fresh Water Resources and International Security." *International Security* 18: 79–112.

This is one of the classic, original articles dealing with the issue of water scarcity. The problem of international security is closely tied to the availability of natural resources, and may be at the heart of many conflicts.

Goble, Dale D. 1994. "Introduction: Public Lands and Agricultural Pollution." *Idaho Law Review* 30: 433.

This special symposium issue from the University of Idaho School of Law examines the role of grazing on public lands, with highly critical assessments of the Bureau of Land Management despite what the author contends is clear congressional intent.

Gonzalez, George A. 2001. "Ideas and State Capacity, or Business Dominance? A Historical Analysis of Grazing on the Public Grasslands." *Studies in American Political Development* 15, no. 2 (fall): 234–244.

This historical overview of grazing on public lands traces key legislation and policies, contending that current regulations benefit the livestock industry, which has controlled federal agencies for decades.

Hatfield, Craig. 1997. "Oil Back on the Global Agenda." *Nature* (May 8): 121.

This brief article examines the ways in which oil drilling has once again become an issue of international proportions, and the potential for conflict.

Hawthorne, Peter. 2000. "Striking at the Root of Civil War." *Time* 155, no. 12 (March 27): 41.

The United Nations has compiled numerous reports on the civil wars that plague Africa and the role of the diamond industry in fostering conflict throughout the continent, especially in Angola.

Kearns, Ethan. 2005. "After Burn: Wildfire Has Destroyed Forests across the West." *American Forests* 111, no 1 (March 22): 26–30.

Rather than dwelling on the damage caused by wildfires, and the way they have changed the country's landscape, the author calls for a massive restoration effort to plant more trees that will cleanse watersheds, provide wildlife habitat, and clean the air.

Kennes, Erik. 2005. "Footnotes to the Mining Story." *Minerals and Energy* 20, no. 1 (March): 23–28.

The author contends that the reason why mining activities and civil war seem to be connected is not because it allows easy access to a country's mining resources but because there are often no other available options for particular resources. He urges companies to work more closely with locally existing networks for more productive outcomes.

Leonard, Bruce. 2005. "Returning the Land to the Giants." *National Parks* 79, no. 1 (winter): 18–22.

After John Muir's visit to the sequoias in 1875, there was public outrage at the possibility that the giant trees might be cut down. Fifteen years later the area was made a national park, and tourist accommodations, such as cabins, stores, and parking lots, began to destroy the trees' delicate root systems. The removal of commercial enterprises from the park did not begin until 1974, at a cost of $70 million.

Limerick, Patricia Nelson. 2005. "Hope and Gloom Out West." *New York Times,* June 22, p. A19.

The author contends that although the West was once seen as the native home of hope because of its natural resources, those same assets are now the second home of tension, conflict, regret, dismay, gloom, and bitterness. The key battlegrounds are over energy production, water, wildfires, and growth.

Lowery, William. 2001. "The Impact of Reinventing Government on State and Federal Parks." *Journal of Policy History* 13, no. 4: 405–428.

This historical view shows how park management has been affected by the need to "reinvent" the relationship between the various levels of government in order to make programs more efficient.

Masland, Tom, et al. 2000. "In Search of Hot Rocks." *Newsweek* 136, no. 2 (July 10): 30–32.

Although much of the attention dealing with the illegal trade in diamonds in Africa has focused on cash sales, this investigative article spotlights the trading of diamonds for guns by rebels of the Revolutionary United Front in Sierra Leone.

Motavalli, Jim. 2005. "Catching the Wind: The World's Fastest Growing Energy Source Is Coming of Age." *E Magazine* (January–February): 26–37.

Offshore wind developments, such as those being built along the East Coast of the United States, are facing stiff opposition from local groups in an uncertain political climate. The author asks whether wind power is a politically acceptable form of renewable energy for the United States.

National Parks Conservation Association. 2005. "Creating Natural Parks." *National Parks* 79, no. 4 (fall): 14–15.

The question and answer format of this article provides an easy-to-understand explanation of how parks are initially proposed and created, the development of park units, and how the appropriations process works in providing funding for national parks.

Nie, Martin. 2004. "A Rule to Sue By." *Headwaters News* (September 30), available at www.headwatersnews.org (accessed January 23, 2006).

The contentious Roadless Rule proposed by the Bush administration would set aside nearly 59 million acres of land from additional road building, an action the author believes will put an enormous responsibility on governors while using rulemaking as a way of solving public lands conflicts.

Nijhuis, Michelle. 2004. "BLM's Crown Jewels Go Begging." *High Country News*, October 25, p. 6.

The National Landscape Conservation Service was intended to serve as an additional source of protection for managing public lands, but the author notes that its role has been steadily declining because of competition over financial resources and jurisdictional battles.

Orogun, Paul. 2004. "'Blood Diamonds' and Africa's Armed Conflicts in the Post–Cold War Era." *World Affairs* 166, no. 3 (winter): 151–161.

The author contends that the end of the Cold War has failed to lead to any peace dividend for the postcolonial African countries. The struggles for control over financial revenues and territories of the diamond industry have led to a catastrophic humanitarian tragedy.

Paskus, Laura. 2005. "The Winds of Change." *High Country News*, May 2, pp. 9–13.

Although many states have enacted a renewable energy portfolio that requires the development of new or expanded sources of power, efforts to implement energy policies are going slowly because of concerns by environmentalists and others over the impact of wind turbines.

Pizer, William A. 2005. "Setting Energy Policy in the Modern Era." *Resources* (winter): 8–10.

As part of a collection of articles, the author contends that resource scarcity is no longer as salient an issue as energy security and environmental challenges that now require government intervention.

Postel, Sandra. 2001. "Growing More Food with Less Water." *Scientific American* 284: 46–51.

While many researchers have proposed ways of dealing with international water scarcity, the author contends that one solution that is being overlooked is finding technology that reduces water loss from evaporation in agricultural production.

Princen, Thomas. 2003. "Principles for Sustainability: From Cooperation and Efficiency to Sufficient." *Global International Politics* 3, no. 1: 33–50.

Although there is considerable scientific research on trends in freshwater availability, there have been thousands of water-related treaties signed, many of which the author believes have shown creativity in dealing with this valuable resource. Cooperation has become the norm, but it is important to consider options such as improving water productivity, regulating groundwater, and preempting conflicts.

Ring, Ray. 2005. "Gold from the Gas Fields." *High Country News*, November 28, pp. 8–13, 19.

Western states are reaping a windfall in tax revenues from oil and gas production, along with thousands of jobs, as part of an energy boom that has both economic and environmental consequences.

Rosenberger, Jack. 2004. "Wasting the West: How Welfare Ranchers and Their Livestock Are Damaging Public Land." *E Magazine* 15, no. 4 (July–August): 20–21.

The San Pedro River area in southeastern Arizona is an example of how a moratorium on livestock grazing on federal lands can benefit the ecology of an area, although the author notes that such success stories are rare because cattle ranchers are heavily subsidized by taxpayers' dollars at well-below-market prices.

Rundle, S. L. 2004. "The Once and Future Federal Grazing Lands." *William and Mary Law Review* 45, no. 4 (March): 1803–1839.

U.S. Supreme Court decisions such as *Public Lands Council v. Babbitt* are typical of the newer conflicts over public lands grazing, especially the changes in regulations under the Clinton administration and unsuccessful attempts at reform to reduce grazing management conflicts.

Schulman, Bruce J. 2005. "Governing Nature, Nurturing Government: Resource Management and the Development of the American State." *Journal of Policy History* 17, no. 4: 375–403.

For a historical background of natural resource management, this article provides a comprehensive analysis of the creation of the administrative bureaucracies that reversed the trend of distributing lands into private hands and instead created a centralized management system in Washington, D.C.

Simon, Bernard. 2002. "Adding Brand Names to Nameless Stones." *New York Times,* June 27, p. W1.

The issue of blood diamonds is causing businesses such as BHP Billiton and De Beers to take a new look at the sources of the gems they use, including new ones in northern Canada from the Ekati mine.

Talheim, Jennifer. 2005. "Report: Public Lands Grazing Costs $123 Million a Year." *Casper Star Tribune* (November 1), at www.casper startribune.net (accessed November 6, 2005).

Examines the results of a Government Accountability Office study showing that grazing fees cover only about one-sixth of the costs of managing the nation's programs that allow livestock to graze on public lands.

Wald, Matthew. 2005. "Power Producers Seek Latest Models of Nuclear Reactors." *New York Times,* March 15, p. F3.

After decades of controversy, power companies are taking a new look at nuclear power as a source of additional energy production, despite public fears over waste and radiation.

Waterman, Ryan. 2001. "Supreme Court Rules against Ranchers, Upholds Grazing Regulations." *Ecology Law Quarterly* 28, no. 2: 552–554.

This 2000 decision preserved the regulatory authority of the Department of the Interior to maintain federal rangeland to both protect the environment and safeguard grazing privileges for ranchers.

Wiebe, Keith, Abebayehu Tegene, and Betsey Kuhn. 1999. "Finding Common Ground on Western Lands." *Rural Development Perspectives* 14, no. 2 (August): 52–56.

Although there are numerous disputes taking place over the ownership of land in the West, the authors contend that voluntary acquisition and conveyance of partial interests in Western land can offer common ground on which to balance competing social, economic, and environmental objectives.

Wolf, Aaron T., et al. 2005. "Water Can Be." *Worldwatch Global Security Brief* 5 (June).

This optimistic view of water scarcity issues downplays doom and gloom scenarios that portray a world in which nations will go to war over water, and argues instead that nations can come together and make rational decisions over international water disputes. Water can be used as a negotiating tool and a communication lifeline connecting countries in time of crisis.

Wuerthner, George. 1991. "How the West Was Eaten." *Wilderness* (spring): 28–37.

An antigrazing activist, Wuerthner examines how the livestock industry has been responsible for the massive degradation of the West, and how political officials have allowed and subsidized cattle grazing on the public lands.

# CD-ROMs/DVD-ROMs/DVDs/Videos

### 2005 Guide to Wildfires and Forest Fires
*Type:*    DVD-ROM
*Date:*    2005
*Source:*    Privately Compiled by Amazon.com.

New technology allows companies like Amazon to develop massive (76,000 pages, in this instance) compilations of government documents. This is an electronic book containing both information, illustrations, and maps that provides a comprehensive resource on wildfires.

### 21st Century Complete Guide to the Healthy Forests Initiative
*Type:*    DVD-ROM

*Date:*    2005
*Source:*    Privately compiled by Amazon.com.

The Bush administration's Healthy Forests Initiative is covered in depth, with additional information about federal wildfire programs, regulations, and policies, along with photographs of fires, profiles of Hotshot crews, and statistics from the National Interagency Fire Center.

### An American Nile
*Type:*    VHS Videocassette
*Length:*    55 minutes
*Date:*    1997
*Source:*    Home Vision Cinema: Public Media Incorporated, Chicago, IL.

Charts the dramatic transformation of the Colorado River from a wild desert waterway, including the construction of Hoover Dam and environmental battles over potential damming of the Grand Canyon.

### Cadillac Desert
*Type:*    VHS Videocassette
*Length:*    55 minutes
*Date:*    1997
*Source:*    Home Vision Cinema: Public Media Incorporated, Chicago, IL.

Traces the fierce political and environmental battles that raged around the transformation of California's Central Valley and the recent trend of diverting water away from agriculture and toward cities and wildlife.

### Can Tropical Forests Be Saved?
*Type:*    VHS Videocassette
*Length:*    120 minutes
*Date:*    1991
*Source:*    Richter Productions.

The demise of tropical rain forest has resulted in barren land covering much of a belt along the equator. Now, thirty-three countries must look for solutions that are cooperative and cost effective.

### End of the Road
*Type:* VHS Videocassette
*Length:* 18 minutes
*Date:* 2000
*Source:* High Plains Films, Missoula, MT.

Short documentary on the half-million miles of roads already constructed in the national forests and the controversy over the nation's roadless rules.

### Fate of the Forest
*Type:* VHS Videocassette
*Length:* 30 minutes
*Date:* 1996
*Source:* Television Trust for the Environment, London, UK.

Indigenous peoples have been at the forefront of developing strategies for preserving tropical rain forests, using their own resources to avoid conflict with loggers and timber companies.

### Fires of the Amazon
*Type:* VHS Videocassette
*Length:* 44 minutes
*Date:* 2002
*Source:* Bullfrog Films, Oley, PA.

Documents the destruction of the Amazon rain forest, explaining the rapid rate of logging and burning, and the growing political clout of local residents.

### The God Squad and the Case of the Northern Spotted Owl
*Type:* VHS Videocassette
*Length:* 57 minutes
*Date:* 2001
*Source:* Bullfrog Films, Oley, PA.

Chronicles the controversial actions of the Endangered Species Committee and proposed timber sales in southwest Oregon that mirror the debate over old growth forests.

### Good Wood
*Type:* VHS Videocassette

*Length:*    45 minutes
*Date:*      1999
*Source:*    Bullfrog Films, Oley, PA.

Asks whether it is possible to stop deforestation while still sustaining communities that depend upon the forest for their livelihood, with coverage of communities in Honduras, British Columbia, and Mexico.

## Grand Canyon

*Type:*      VHS Videocassette
*Length:*    95 minutes
*Date:*      2002
*Source:*    Firstlight Pictures, Maplewood, NJ.

Documents the human history of the Grand Canyon, including discovery and exploration, environmental conditions, recreational use, and history.

## The Great Forest

*Type:*      VHS Videocassette
*Length:*    104 minutes
*Date:*      2003
*Source:*    High Plains Films, Missoula, MT.

Identifies the conflict between conservationists and the wood products industry in three short documentaries that examine the loss of virgin forests in the East.

## The Greatest Good

*Type:*      VHS Videocassette
*Length:*    120 minutes
*Date:*      2005
*Source:*    U.S. Forest Service, Washington, DC.

U.S. Forest Service centennial film traces the agency's history throughout the twentieth century, examining conflicts over timber, grazing, fire, wilderness, and recreation.

## Hoover Dam

*Type:*      VHS Videocassette
*Length:*    60 minutes
*Date:*      1999
*Source:*    PBS Video, Alexandria, VA.

Describes the making of a national monument and the environmental conflicts involved, using archival footage and photographs.

## Human Faces behind the Rain Forest

*Type:* VHS Videocassette
*Length:* 30 minutes
*Date:* 2001
*Source:* First Run Icarus Films, Brooklyn, NY.

Documents the dramatic events surrounding the harvest of the opium poppy crop in the Colombian rain forest through the experiences of the indigenous peoples involved.

## Last Oasis

*Type:* VHS Videocassette
*Length:* 55 minutes
*Date:* 1997
*Source:* Home Vision Cinema: Public Media Incorporated, Chicago, IL.

The story of how America's large dams became examples for water projects in developing countries, and the conflicts that have arisen in those nations.

## The Last Stand: Ancient Redwoods and the Bottom Line

*Type:* VHS Videocassette
*Length:* 57 minutes
*Date:* 2002
*Source:* University of California Extension Center for Media, Berkeley, CA.

Examines the destruction of the ancient redwoods in northern California, with testimony from economists, scientists, forest activists, and local residents.

## Logs, Lies, and Videotape

*Type:* VHS Videocassette
*Length:* 12 minutes
*Date:* 1996
*Source:* Green Fire Productions, Eugene, OR.

Looks at the impact of a logging measure passed by Congress that directs the Forest Service and Bureau of Land Management to accelerate salvage logging.

## Motor

Type: VHS Videocassette
Length: 38 minutes
Date: 1999
Source: High Plains Films, Missoula, MT.

Shows how recreational use by off-highway vehicles, personal watercraft, and other motorized vehicles is affecting wilderness areas and the nation's lakes, deserts, and forests.

## Mulholland's Dream

Type: VHS Videocassette
Length: 85 minutes
Date: 1997
Source: Home Vision Cinema: Public Media Incorporated, Chicago, IL.

Tells of William Mulholland's search for water for the people of Los Angeles and the building of the aqueduct 250 miles from the Owens Valley to Southern California.

## Oil on Ice

Type: VHS Videocassette
Length: 90 minutes
Date: 2004
Source: Sierra Club Productions, San Francisco, CA.

Investigates the possible environmental and cultural effects of proposed drilling in Alaska's Arctic National Wildlife Refuge. The film emphasizes the impact on the native Gwich'in people, caribou herds, and global climate.

## Powder River Country

Type: VHS Videocassette
Length: 34 minutes
Date: 2005
Source: High Plains Films, Missoula, MT.

Shows the transformation of Wyoming's Bighorn Mountains area, which is a potential new source of natural gas, during a time of energy debates.

## Razing Appalachia

Type: VHS Videocassette

*Length:*   54 minutes
*Date:*   2003
*Source:*   Bullfrog Films, Oley, PA.

Portrays the struggle of the people of Blair, West Virginia, against a mining company whose jobs in local coal mines are a major source of employment for the region.

### Restoring the Everglades
*Type:*   VHS Videocassette
*Length:*   15 minutes
*Date:*   1998
*Source:*   National Parks and Conservation Association, Washington, DC.

Reviews issues of wetland conservation, agricultural pollution, and efforts to restore the national park.

### Save the Sungmi Mountain
*Type:*   VHS Videocassette
*Length:*   36 minutes
*Date:*   2003
*Source:*   Diffusion Films, Korea.

Documents the efforts of the municipal government in Seoul, Korea, to build a reservoir on top of a mountain, and the struggle between politicians and villagers who oppose the project.

### Since the Company Came
*Type:*   VHS Videocassette
*Length:*   52 minutes
*Date:*   2000
*Source:*   First Run Icarus Films, Brooklyn, NY.

Story of a remote village in the Solomon Islands that invites a Malaysian company to log their tribal land, and then faces disputes over logging royalties and the preservation of forests and native traditions.

### Taking Stock: Living with Change in the American Southwest
*Type:*   VHS videocassette
*Length:*   52 minutes
*Date:*   2004
*Source:*   Grand Canyon Association, Flagstaff, AZ.

Covers a presentation by Colorado Plateau experts Rose Houk and Michael Collier about the nature of ecological change in the Southwest, focusing on public lands, water, and conservation issues.

### Tales of the San Joaquin

*Type:*      VHS Videocassette
*Length:*    27 minutes
*Date:*      2004
*Source:*    Christopher Beaver Films, Sausalito, CA.

Explores this California river's history and the issues of water rights that developed after the building of the Friant Dam in the 1940s.

### Thirst

*Type:*      DVD
*Length:*    62 minutes
*Date:*      2004
*Source:*    Bullfrog Films, Oley, PA.

Without narrative, shows how the debate over water rights between communities and corporations can serve as the catalyst for resistance to globalization.

### Treasuring Our Natural Heritage: Biodiversity Science Games

*Type:*      CD-ROM
*Date:*      2003
*Source:*    Boise, Idaho Public Television

Computer games that deal with natural resources, human ecology, and biological diversity, such as Habitat Hike, Nature's Kitchen, and Linked to the Land.

### Trees Are the Answer

*Type:*      VHS Videocassette
*Length:*    30 minutes
*Date:*      2000
*Source:*    Greenspirit, Ltd., Vancouver, BC.

Based on the book of the same name, this controversial film contends the environmental movement's current thinking about forests is on the wrong track. The film proposes that rather than reducing our consumption of wood, we should be planting more trees and using more renewable wood to reduce our reliance on non-renewable fuel sources.

### Tree Sit: The Art of Resistance

| | |
|---|---|
| *Type:* | VHS Videocassette |
| *Length:* | 120 minutes |
| *Date:* | 1998 |
| *Source:* | Earth Films, Redway, CA. |

In documentary format, this film chronicles the protests of the Redwood Summer that were aimed at stopping logging in the Headwaters Forest, one of the last stands of old growth forest in the United States.

### Trinkets and Beads

| | |
|---|---|
| *Type:* | VHS Videocassette |
| *Length:* | 53 minutes |
| *Date:* | 1996 |
| *Source:* | First Run Icarus Films, Brooklyn, NY. |

Documents the lives of the Huaorani, a small tribe of Ecuadorian Indians, who after pressure from foreign oil companies, agreed to allow oil drilling on their land.

### Troubled Waters: The Dilemmas of Dams

| | |
|---|---|
| *Type:* | VHS Videocassette |
| *Length:* | 53 minutes |
| *Date:* | 2003 |
| *Source:* | The Video Project, San Francisco, CA. |

Looks at the controversies over dams from the perspectives of environmental, cultural, economic, and spiritual issues, using personal interviews and archival footage.

### Water for the Fields

| | |
|---|---|
| *Type:* | VHS Videocassette |
| *Length:* | 27 minutes |
| *Date:* | 2003 |
| *Source:* | DW-TV, Germany. |

Explores the issue of agricultural irrigation throughout the world, showing innovative ideas and the destruction caused by deforestation.

## Water, Land, People and Conflict

*Type:* VHS Videocassette
*Length:* 29 minutes
*Date:* 1998
*Source:* Center for Defense Information, Washington, DC.

Argues that the idea of a healthy environment is just as vital to national security as military strength.

## Whose Home on the Range?

*Type:* VHS Videocassette
*Length:* 55 minutes
*Date:* 1999
*Source:* Bullfrog Films, Oley, PA.

Covers the opposition to federal land management policies in Catron County, New Mexico, and conflicts among ranchers, loggers, environmental groups, and the U.S. Forest Service.

## Wildland

*Type:* VHS Videocassette
*Length:* 35 minutes
*Date:* 2000
*Source:* High Plains Films, Missoula, MT.

Expresses the importance of retaining untouched wilderness, featuring the nation's most prominent natural spaces.

## Wind River

*Type:* VHS Videocassette
*Length:* 34 minutes
*Date:* 1999
*Source:* High Plains Films, Missoula, MT.

Describes how water rights are awarded to Wyoming farmers on a seniority basis, examining core community values and concerns over fairness.

# Glossary

**Acre:** An acre of land measures about 43,560 square feet, with a square, one-acre plot measuring about 209 feet by 209 feet.

**Acre Foot:** The amount of water needed to cover an acre of land, one foot deep. An acre foot is equivalent to 325,851 gallons of water.

**Adaptive Management:** A process for adjusting management and research decisions to better achieve management objectives, recognizing that knowledge about natural resource systems is uncertain.

**Alternative Energy:** Energy that is not popularly used and is usually environmentally sound, such as solar or wind energy.

**Alternative Fuels:** Transportation fuels other than gasoline or diesel, including natural gas, electricity, methanol, and biofuels.

**Ancient Forest:** A forest that is typically older than 200 years with large trees, dense canopies, and an abundance of diverse wildlife.

**ANWR:** Alaska National Wildlife Refuge. A controversial protected area that is under consideration as a source of oil for the United States.

**Aquifer:** An underground source of water, often pooled beneath the surface.

**AUM:** Animal Unit Month. The amount of forage that a cow and her calf can eat in one month, used by the U.S. government to calculate the federal grazing fee.

**Biodiversity:** A large number and wide range of species and animals, plants, fungi, and microorganisms. Ecologically, wide biodiversity is conducive to the development of all species.

**Biomass:** The total woody material in a forest, both merchantable and material, such as small branches and leaves, that results from a logging operation. Biomass can also consist of other types of vegetation that can be burned, including rotting fruits and vegetables, lawn clippings, and landscape debris. Some types of biomass are used as fuel.

**Biosphere:** The part of the earth and atmosphere in which living organisms exist or that is capable of supporting life.

**Biosphere Reserve:** A part of an international network of preserved areas designated by the UN Educational, Scientific, and Cultural Organization (UNESCO). Biosphere reserves are vital centers of biodiversity in which research and monitoring activities are conducted to preserve healthy natural systems threatened by development.

**Blood Diamonds** (see conflict diamonds).

**Board Foot:** A unit of wood measuring 144 cubic inches, used as a measurement in timber harvesting.

**Boreal Forests:** The forests found in the areas of the far north latitudes, below the treeless tundra of the polar region. Boreal forests are the biggest terrestrial ecosystem in the world, almost completely intact, free of roads and industrial development.

**Bourse:** A stock exchange–type institution that is typically a corporation that provides for trade through a centralized system of pooled goods.

**BTU:** British thermal unit. A unit of energy used primarily in the United States, calculated as the amount of heat required to raise the temperature of one pound of water by one degree Fahrenheit.

**Carbon Sinks:** Areas where carbon is stored and not released into the atmosphere, such as forests and wetlands.

**Clear-cutting:** In logging, the practice of cutting down all the trees in a designated parcel of land, leaving only the stumps or other debris.

**Collaboration:** A process involving two or more groups or interests who agree to cooperate to solve an identified problem.

**Commodity:** Treating a resource as if it were a crop available for harvest, such as trees considered as a product of the forest.

**Conflict Diamonds:** Diamonds, in either rough or polished form, extracted from or traded in areas where there is ongoing armed warfare, such as Sierra Leone or Liberia.

**Crude Oil:** Composed of natural gas liquids, refinery feedstocks, and additives as well as other hydrocarbons.

**Custodial Management:** The concept that the government is not just the owner of land but also the steward of it for future use.

**Ecology:** A branch of science concerned with the interrelationship of organisms and their environment.

**Ecoregion:** An area defined by environmental conditions and natural features, or defined by its ecology.

**Ecosystem:** All of the living organisms (animals, plants, and microbes) in a given area, and the processes of life cycles, usually described by the predominant form of life, such as a marine ecosystem.

**Ecoterrorism:** The intentional sabotage or destruction of buildings, vehicles, or activities by persons or groups opposed to a controversial environmental project, agency, or company.

**Endangered Species:** In generic use, plants or animals in danger of going extinct; in the United States the term is applied under the Endangered Species Act to species when the total number remaining may not be sufficient to reproduce enough offspring to guarantee survival of the species.

**Extinction:** The state of a species no longer existing throughout its entire range.

**First Nations:** The original people inhabiting an area, often synonymous with Indian, aboriginal, or indigenous people. The term is often applied to the governments, although there is no legal standing for these groups under international law.

**Forage:** Vegetation such as grasses, small shrubs, and other plants used for grazing by livestock or wildlife, or the act of eating these types of plants.

**Geographic Information System:** Technology that provides data on a specific plot of land or space, usually through satellite imagery.

**Grazing Fee:** An amount charged by the government or by a private owner to allow a rancher to let livestock forage on designated parcels of land.

**Gross Domestic Product:** A measure of the size of a nation's economy, defined as the market value of all final goods and services produced within a country in a given period of time.

**Groundwater:** Water that accumulates below the surface in pools or aquifers.

**Habitat:** The natural area in which plants and animals grow and reproduce.

**Headright:** A policy allowing an immigrant 50 acres of land upon settling it himself, used to encourage colonization.

**Hectare:** A unit of area equal to 10,000 square meters, commonly used for measuring land areas in the fields of agriculture, forestry, and public planning.

**Homesteading:** The acquisition of up to 160 acres of land by any head of a family over twenty-one years of age, upon proof that the person had resided upon and cultivated the land for at least five years.

**Hydrology:** The scientific study of the properties, distribution, and effects of water on the earth's surface, in the soil and underlying rocks, and in the atmosphere.

**Indigenous:** Native to a region or place. There are an estimated 370 million people considered indigenous to about seventy countries worldwide. Indigenous people are the inheritors and practitioners of unique cultures and ways of relating to other people and the environment.

**Kimberley Process:** An international initiative established to develop practical approaches to tracing the origin of diamonds through certification and regulations, named after Kimberley, South Africa, where the first major conference on diamond certification was held.

**Multiple Use:** Term coined by Gifford Pinchot to refer to the management of land for more than one purpose, including recreation, commercial production, or aesthetics.

**Natural Resources:** The land, water, and atmosphere, and the components upon or within them, such as minerals, and animal and plant life.

**NGO:** Nongovernmental organization, or a group not affiliated with a formal institution of government, such as a trade association, charitable group, or collective interest.

**Old Growth:** Forests that are distinguished by old trees representing the last stage of stand development, with age dependent on the type of vegetation and climate.

**OPEC:** Organization of Petroleum Exporting Countries, made up of eleven member countries.

**Organic Act:** The initial legislation used to establish an agency within the U.S. government.

**PETA:** People for the Ethical Treatment of Animals, an animal rights organization.

**Plenary:** A business meeting or session held during a conference.

**Problem Displacement:** Describes a situation in which a problem is passed along to another party, rather than seeking a solution to the problem.

**Public Domain:** Lands acquired by the United States from other countries through purchase or other forms of acquisition.

**Rangeland:** Areas of vegetation where livestock or wildlife graze openly, including shrub land and some barren areas. Rangeland is usually considered less valuable than farmland.

**RARE:** Roadless Area Review Evaluation. The U.S. Forest Service conducted an evaluation of wilderness areas to determine which parcels should be declared off-limits to road building through two inventories in the late 1970s.

**Reservation:** Under Article IV of the U.S. Constitution, Congress has the power to reclassify a given area of land from being available for sale or other form of disposal to one in which it is committed to a specific public use. The first major reservation occurred in 1872, when land for Yellowstone National Park was set aside by Congress.

**Reserve:** An area that has been designated for a specific use, such as a forest reserve or a wildlife reserve. It may also be afforded some kind of protected status.

**Riparian:** The areas adjacent to a stream, river, lake, or wetland that are distinct from the water and uplands, such as the banks of a river or its flood plain.

**Rough:** Diamonds in rocks before they are converted into polished stones.

**Sagebrush Rebellion:** A movement started by ranchers and miners during the late 1970s in response to efforts of the federal government to improve the management of public lands.

**Salvage Logging:** The logging of dead or diseased trees in order to improve overall forest health.

**Scheme:** A systematic way of accomplishing a task; a mechanism for implementing a policy.

**Second Growth Forests:** Forests that have grown back after being logged.

**Slash:** The leafy debris that is left on the ground after a forest is logged. Slash is usually gathered into piles and then burned, rather than being used as biomass fuel.

**Soot:** A fine, sticky powder, composed mostly of carbon, formed by the burning of fossil fuels.

**Species:** A group of related organisms having common characteristics.

**Squatter:** A person who obtained possession of land simply by staying on it without formal title.

**Sulfur Dioxide:** A heavy, odorous gas that can be condensed into a clear liquid. It is used to make sulfuric acid, bleaching agents, preservatives, and refrigerant, and is a major contributor to acid rain in industrial areas.

**Surface Water:** Water located above the ground, usually in rivers, lakes, and other bodies of water.

**Sustainability:** Maintenance of the resource to produce or maintain a relatively constant supply or number, such as sustainable agriculture.

**Tenure:** In land use, ownership conferred upon an individual on the basis of military service, or issued by a legal authority.

**Third World:** Countries that are in the developing stage of building an economy infrastructure.

**Threatened Species:** In the United States, under the Endangered Species Act, a plant or animal that is not yet endangered but that requires protection to reduce the potential for extinction.

**Tree Spiking:** The process of hammering nails into trees about to be cut down as a way of ruining the saw blades, potentially injuring loggers. The practice is alleged to be one of the major tactics used by groups associated with ecoterrorism and sabotage, such as Earth First!

# Index

# About the Author

Jacqueline Vaughn is professor of political science at Northern Arizona University, where she specializes in public policy and administration. Professor Vaughn holds a Ph.D. in political science from the University of California, Berkeley, where she also attended the Goldman School of Public Policy. She taught previously at the University of Redlands and at Southern Oregon University. Professor Vaughn has a broad spectrum of academic experience in both the public and private sectors. Her environmental background stems from her work with the South Coast Air Quality Management District in southern California, and with Southern California Edison, where she served as a policy analyst. Professor Vaughn's previously published works include *The Play of Power: An Introduction to American Government; Green Backlash: The History and Politics of Environmental Opposition in the United States; Disabled Rights: American Disability Policy and the Fight for Equality; Environmental Activism: A Reference Handbook; George W. Bush's Healthy Forests: Reframing the Environmental Debate* (coauthored with Hanna J. Cortner); and *Environmental Politics: Domestic and Global Dimensions,* now in its fifth edition.